Industrial Pneumatic Control

FLUID POWER AND CONTROL

A Series of Textbooks and Reference Books

1. Hydraulic Pumps and Motors: Selection and Application for Hydraulic Power Control Systems, *by Raymond P. Lambeck*
2. Designing Pneumatic Control Circuits: Efficient Techniques for Practical Application, *by Bruce E. McCord*
3. Fluid Power Troubleshooting, *by Anton H. Hehn*
4. Hydraulic Valves and Controls: Selection and Application, *by John J. Pippenger*
5. Fluid Power Design Handbook, *by Frank Yeaple*
6. Industrial Pneumatic Control, *by Z. J. Lansky and Lawrence F. Schrader, Jr.*

Other Volumes in Preparation

Industrial Pneumatic Control

Z. J. Lansky
Lawrence F. Schrader, Jr.

Parker Hannifin Corporation
Cleveland, Ohio

Marcel Dekker, Inc. New York and Basel

Library of Congress Cataloging in Publication Data

Lansky, Z. J. (Zdenek J.), [date]
Industrial pneumatic control.

(Fluid power and control ; 6)
Includes index.
1. Pneumatic control. I. Schrader, Lawrence F., [date] II. Title.
TJ219.L35 1986 629.8'045 85-29308
ISBN 0-8247-7494-9

MARCEL DEKKER, INC.
270 Madison Avenue, New York, New York 10016

Current printing (last digit):
10 9 8 7 6 5 4 3 2 1

Printed in the United States of America

Preface

This text, like most other technically oriented works, is the result of the efforts of many individuals. A broad range of talents is represented, including those of engineers, designers, technicians, and maintenance personnel. It is through the free communication among experts in these different disciplines that this text has been made possible. In addition, the text brings together many insights and experiences that have been described in numerous technical publications of the Parker Hannifin Corporation.

This book was written for those concerned with learning about the pneumatic directional control valve and regulator, and pneumatic circuitry. It touches on practically all of the typical problems encountered by designers, engineers, installers, and maintenance personnel who are involved with pneumatic systems in one way or another. Emphasis has been placed on component construction and function, as well as the installation, maintenance, and troubleshooting of malfunctioning components. However, the text also contains much technical information that will be useful to engineers in their selection of pneumatic system components and pneumatic circuitry.

The first three chapters discuss the basics of valve types, functions, and construction. The fourth and fifth chapters cover various pneumatic circuits and valve applications. Chapter 6 describes useful procedures for sizing pneumatic directional control valves and pressure regulators. The next two chapters are designed to aid those responsible for keeping the system functioning. These two chapters will help facilitate installation, maintenance, trouble-shooting, and repair of air valves. Finally, Chapter 9 gives the reader insight into innovations taking place in the field of pneumatics, and suggests how these innovations can be applied to solve circuit problems.

The material presented here will prove useful to plant engineers, design engineers, and maintenance personnel in their day-to-day problems. Those new to the field of fluid power can take advantage of the years of experience contained between the covers of this text so as to avoid the common pitfalls.

The thoughts and ideas of many people are included in this volume. Without their contributions, much of the material would not be available here. Their efforts have made it possible to present what we hope is a useful reference work.

We thank the staff of Marcel Dekker, Inc., who encouraged us to prepare this text.

All figures in this volume are courtesy of the Parker Hannifin Corporation unless otherwise noted.

Z. J. Lansky
Lawrence F. Schrader, Jr.

As difficult as it would be to acknowledge all those who contributed to this work, it is felt that special thanks should go to Ms. Cathy Miller, who spent many hours typing this manuscript, and to Mrs. L. Schrader, for the many hours she spent proofreading the manuscript and without whose constant support and encouragement this text would not have come to be.

L.F.S.

Contents

Industrial Pneumatic Control

1

Control of Pneumatic Energy

1.1. INTRODUCTION

The development of mass production on assembly lines as a standard industrial process increased the demand for and possible applications of compressed air. A list of typical industrial applications is presented in Figure 1.1.

Air Brake	Conveying	Hoisting
Air Cylinder	Die Casting	Mixing
Air Motor	Drilling	Paint Spray
Atomizing	Elevating	Pile Driving
Buffing	Forming	Pressurizing
Chipping	Grinding	Process Control
Reaming	Riveting	Stapling
Screw Driving	Transferring	Clamping

Figure 1.1 Industrial applications of compressed air.

In recent years, compressed air has been applied to control circuitry, dental drills, and surgical instruments, as well as to many industrial processes requiring high forces or impact blows. Lightweight, durable, and safe pneumatic tools, such as pneumatic staplers and pneumatically powered impact wrenches, are familiar to many. Other modern applications of the technology, which began with the lung power blowgun, are as varied as compressed air starters for diesel motors, snowmaking machines for ski slopes, and pneumatic lifts in automobile service stations.

This chapter will introduce the basic components found in the typical industrial pneumatic (compressed air) system. The remainder of the text will concentrate on one of these components, the pneumatic directional control valve.

The energy of the compressed air that is transmitted through a pneumatic system must be directed and under complete control at all times. If it is not under control, useful work may not be done and machinery or machine operators might be harmed. One of the advantages of transmitting energy pneumatically is that the energy can be controlled relatively easily.

1.2. VALVES

The basic pneumatic valve is a mechanical device consisting of a body and a moving part which connects and disconnects passages within the body. The flow passages in pneumatic valves carry air under pressure. The action of the moving part may control system pressure, direction of flow, and rate of flow. Let us briefly examine each of these topics, beginning with pressure.

1.3. CONTROL OF PRESSURE

Pressure in a pneumatic system must be controlled at several points—at the compressor and after the air receiver tank as well as at the point of usage (Fig. 1.2). Control of pressure is required at the compressor to protect the compressor, the general storage, and the compressed air transmission system. Control of pressure at the point of air usage is necessary for functional safety and so that actuators receive the proper pressure and energy is not wasted.

1.3.1. Control of Pressure After a Compressor

Characteristically in a pneumatic system, the energy delivered by a compressor is not used immediately, but is stored as potential energy in an air receiver tank and the connected piping system in the form of compressed air. In most applications, a compressor is designed into a system so that it operates intermittently. A compressor usually delivers compressed air to a receiver tank until the desired high pressure is reached, and then its air delivery is controlled. In most large compressors, this pressure control function is provided by unloading the compressor, either by closing the compressor intake valve or through some other means which prevents the compression of air. In smaller units, the control function results from the pressure level sensed in or near the receiver; when air pressure in the tank decreases, the compressor delivery is increased (within its design limits). Compressor operation in this

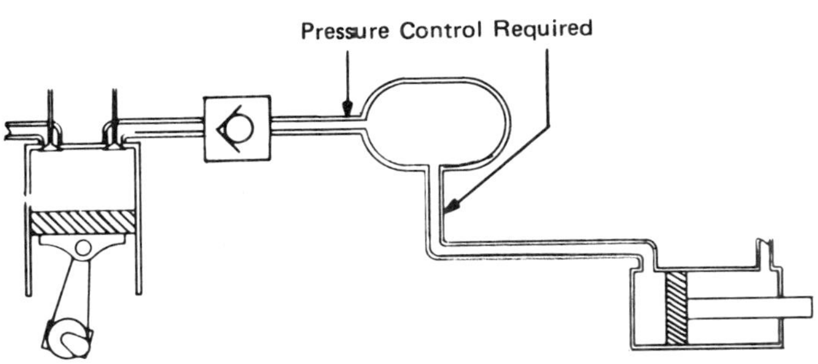

Figure 1.2

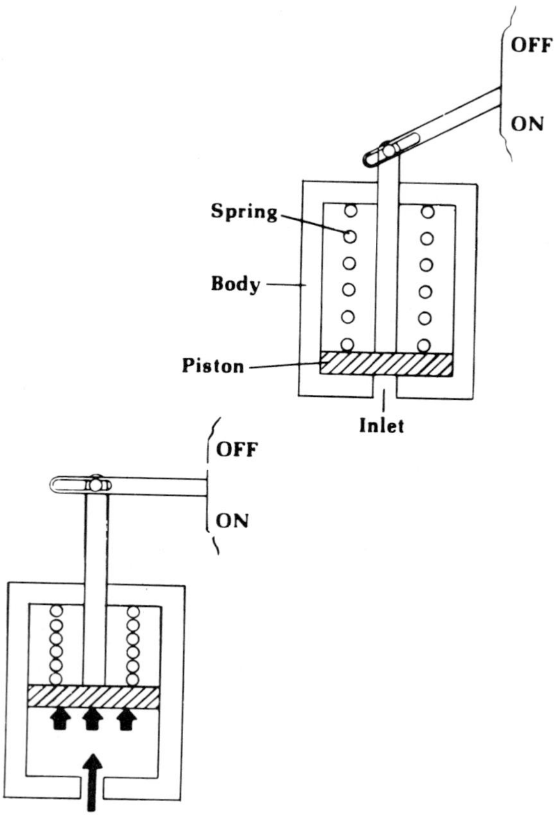

Figure 1.3

manner is a power savings for the system. One way to accomplish this function is through the use of a pressure switch (Fig. 1.3).

A common way of sensing tank pressure and controlling actuation and deactuation of a relatively small (5-10 hp) compressor is with a pressure switch. Large compressors use other means of various types.

In a pressure switch connected to a system, pressure is sensed at the bottom of the piston through the pressure switch inlet. When the pressure being sensed is at its low level, the spring pushes the piston down. In this position, it makes a contact that causes an electrical signal to turn on the compressor drive.

As pressure in the receiver tank rises, it is also sensed by the pressure switch, where the pressure forces the piston upward. With system pressure at its high level, the piston breaks the electrical contact, shutting down the compressor driving motor. However, it should be noted that in large compressors, the compressor is not shut off. Instead, it is unloaded in some way to prevent the compression stroke from compressing any air and thus requiring large amounts of power.

Maximum pressure developed by a compressor is designed to be regulated by a control system which senses discharge or tank pressure. In case of an emergency, as in the failure of a control system to function properly, a positive displacement compressor system should have a safety relief valve (Fig. 1.4). This is mandated by law for most applications.

A safety relief valve consists of a valve body with inlet and exhaust passages and an internal moving part. In the illustration, the moving part is a poppet, which is held biased in one position by a spring. To comply with safety codes the valve will have a manual test lever, which should be actuated periodically to be certain that it is free to work properly.

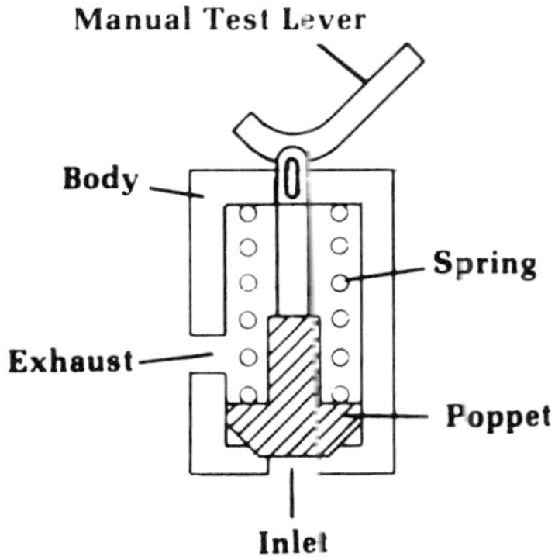

Figure 1.4

The safety relief valve is a normally nonpassing (closed) one-way directional control valve. The poppet of the safety relief valve is seated on the valve inlet. A spring holds the poppet firmly on its seat. Air cannot pass through the valve until the force of the spring biasing the poppet is overcome. Air pressure at compressor's outlet is sensed directly on the bottom of the poppet. When the air pressure reaches the preset control level, the force of the air on the poppet is greater than the spring force. When this happens the spring will be compressed, the poppet will move off its seat, and air will exhaust through the valve vent (exhaust) ports (Fig. 1.5).

Safety relief valves on a compressor are not intended to be operated frequently; they are safety devices. Frequently, safety relief valves are equipped with whistles or horns to alert personnel that something has failed, and that a problem or unsafe condition exists. Good maintenance and safety procedures require the periodic checking of safety valves to verify that they can move freely. This can often be done by moving the manual test lever to open the valve.

1.3.2. Pressure Regulator

Next we must control pressure downstream of the receiver tank. In a pneumatic system, energy which will be used by the system and transmitted through the system is stored as potential energy in an air receiver tank in the form of compressed air. A pressure regulator is positioned after a receiver tank and is used to control

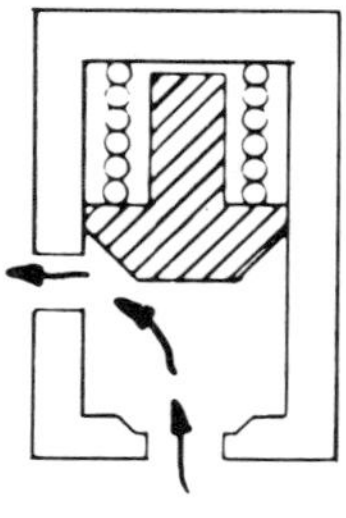

Figure 1.5

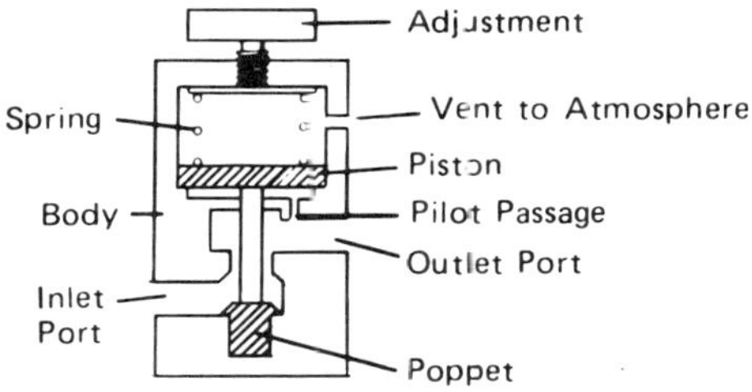

Figure 1.6

the pressure of this stored energy to a branch of a circuit. This valve can also be used as a directional control valve. This will be discussed later in the text. (See Fig. 1.6.)

A pressure regulator valve consists of a valve body with inlet ("primary") and outlet ("secondary") passages. The regulator has a moving part which controls the size of the opening between the primary and secondary passages. In our illustration, the movable part is a poppet which is connected to a piston. This piston is biased away from its seat ("open") by an adjustable spring.

The pressure regulator is a normally passing (open) valve (Fig. 1.7). That is, the flow passage normally allows air to flow freely. With a regulator positioned after a receiver tank, air from the receiver can expand (flow) through the valve to a point downstream of the secondary passage. When the pressure in this passage of the regulator rises, it is transmitted through an internal pilot passage which leads to the piston area on the opposite side of the spring.

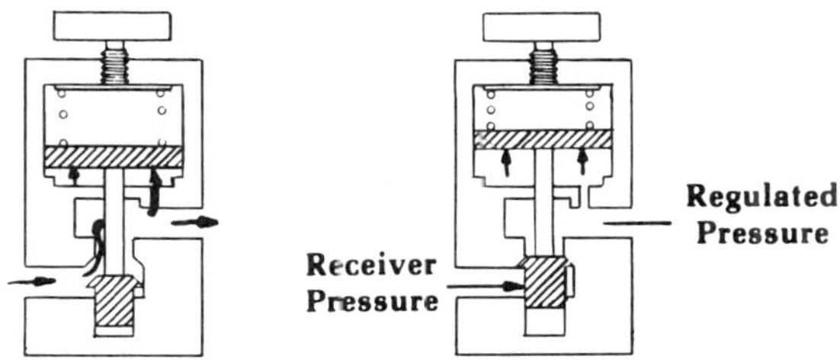

Figure 1.7

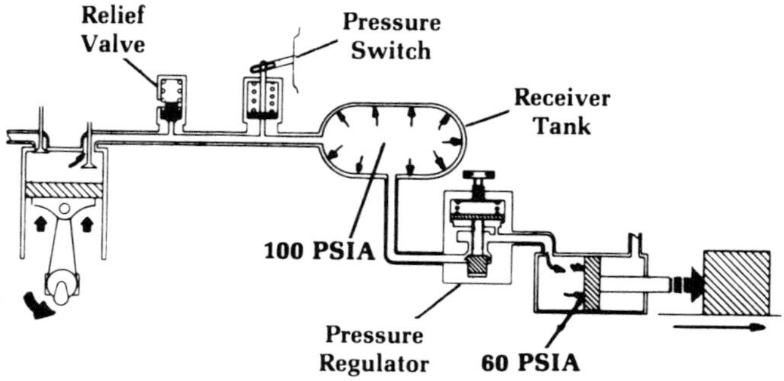

Figure 1.8

The piston has a relatively large surface area exposed to secondary pressure and is therefore responsive to secondary pressure fluctuations. When the controlled pressure nears the present level, the piston moves upward, allowing the poppet to move toward its seat to modulate (or throttle) the flow. The poppet blocks the flow once it seats and does not allow pressure to continue building downstream. In this way, air at a controlled pressure is made available to an actuator downstream. Also, the use of a pressure regulator ensures that energy in an air system is not wasted. For instance, if an air receiver charged to 100 psia were allowed to transmit its full pressure to an actuator which required only 60 psia to perform its useful work function, pressure energy would be wasted. This may be avoided by placing, before the actuator, a regulator that is set to deliver only the amount of pressure needed to overcome load resistance plus a few additional pounds per square inch to develop flow (Fig. 1.8).

1.4. CONTROL OF FLOW RATE

In a pneumatic system, actuator speed is determined by how quickly the actuator can be filled and exhausted of air. In other words, the speed of a pneumatic actuator depends on the force available from the pressures acting on both sides of the piston, a result of the volume flow (cubic feet per minute) into the inlet and out of the exhaust port.

We have seen that a regulator will influence actuator speed by portioning out to its leg of a circuit the pressure required to equal the load resistance at an actuator. This additional pressure is used to develop air flow. Even though this is the case, pressure regulators are not used to vary actuator speed. In a pneumatic system, actuator speed is affected by a restriction, as with a needle valve or a needle valve with a bypass check, often called a "flow control" valve. This valve will meter flow at a constant rate only if the pressure at its inlet and outlet are constant. This can only be achieved if the system resistance and the total load (friction, load) are constant and if the pressure at the inlet of the cylinder and the pressure at its outlet do not vary throughout the entire work cycle. This valve does not control flow, it only affects flow. Speed is affected by other factors. For example, if the actuator encountered an added resistance because of friction and/or load changes, its speed would decrease or the actuator would stop until the effective pressure changed sufficiently to restart the actuator.

> *Safety Note:* Because of the compressibility of air, "speed control" circuits must be examined very closely for all possible conditions that would cause force—and thus velocity—changes. If an actuator paused because of such changes, there would be a very rapid increase in speed once the resistance was overcome. This would be a very dangerous condition for the operator and other nearby personnel, as well as for the function or equipment.

1.4.1. Needle Valves

As indicated earlier, a needle valve in a pneumatic system affects the operation by causing a restriction. The typical needle valve consists of a valve body and an adjustable part (Fig. 1.9). The ad-

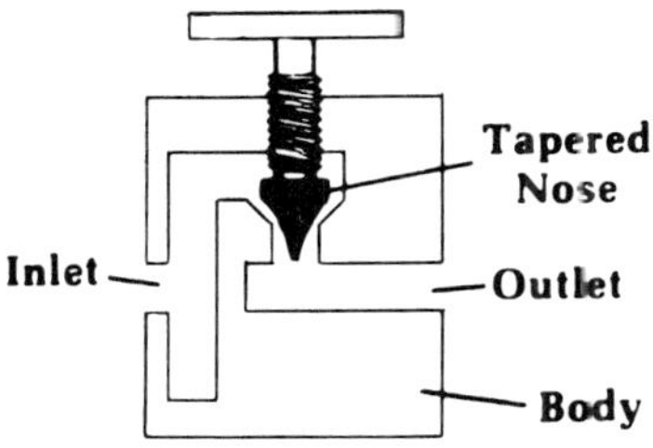

Figure 1.9

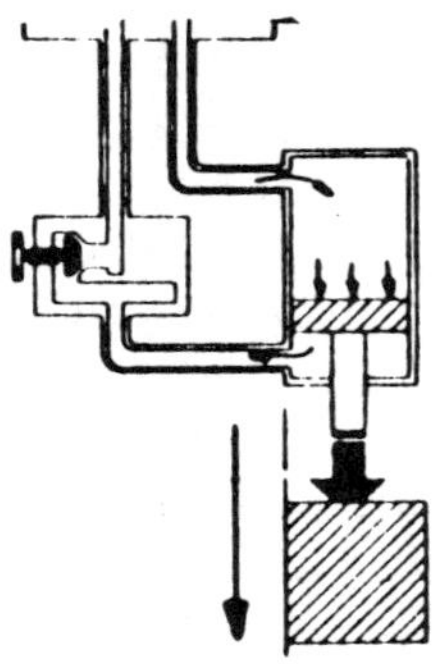

Figure 1.10

justable part in our example is a tapered-nose rod which is threaded into the valve body. The part in this valve is described as adjustable because it can be repositioned or "adjusted" fairly easily.

In order to reduce the speed of an actuator in a pneumatic system, a needle valve is sometimes positioned in a circuit so that it restricts air flow exhausting from an actuator port (Fig. 1.10). The more the tapered-nose rod is screwed toward its seat in the valve body, the greater the restriction of free flow.

By restricting the exhaust air flow in this manner, a back pressure is generated within the actuator, thus reducing the forces available to create motion. This means that a larger portion of regulator pressure is used to overcome the resistance at the actuator and less pressure energy is available to develop flow.

With fewer cubic feet per minute flowing into the actuator, actuator speed decreases.

By controlling the amount of restriction developed by a needle valve is used to change the direction of air flow to and from the actotal load sensed by the actuator is constant.

1.5. CONTROL OF ACTUATOR DIRECTION

Once a cylinder is extended (or a rotary actuator has moved to its limit), it must be retracted or reversed. To perform this task, a valve is used to change the direction of air flow to and from the actuator. For linear actuation, this is generally done by using a double-acting cylinder and one or two directional control valves.

1.5.1. Double-Acting Cylinder

The double-acting cylinder has a port at each end of the cylinder body by which air under pressure can enter or exhaust (Fig. 1.11). This causes the piston rod to extend or retract (double-acting). To distinguish the ports on a double-acting cylinder, we will label one A and the other B. (The motion of rotary actuators is controlled in a similar manner. It should be recognized that, unless otherwise stated, the control of a rotary actuator is similar to that of a linear actuator cylinder.)

1.5.2. Directional Control Valve

Next, to change the direction of air flow to and from the actuator we use a directional control valve. The moving part in a directional control valve will connect and disconnect internal flow passages within the valve body. This action results in a control of air flow direction (Fig. 1.12).

The typical directional control valve consists of a valve body with four internal flow passages and a moving part, a spool, which alternately connects an actuator port to supply pressure or to exhaust.

With the spool in one extreme position, supply pressure is connected to port B and port A is connected to exhaust. If actuator port A is connected to the valve port A and actuator port B to the valve port B, then with the spool in the other extreme position, supply pressure is connected to actuator passage A and actuator

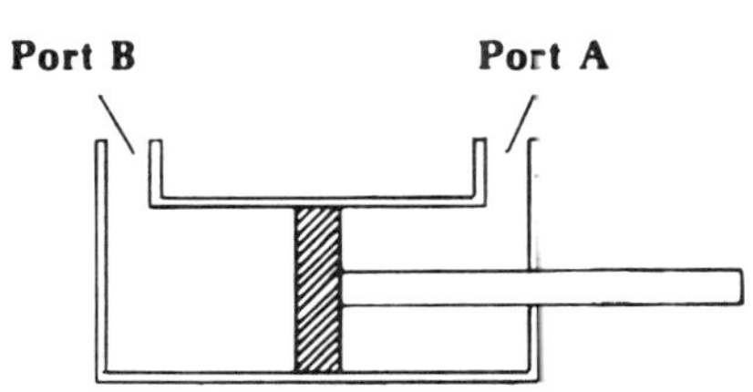

Figure 1.11

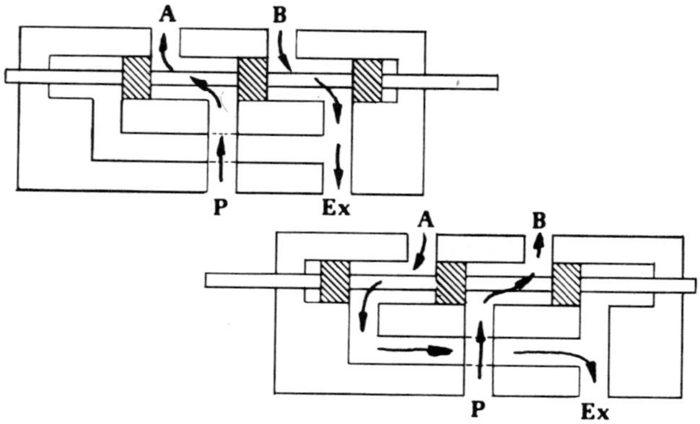

Figure 1.12

passage B is connected to exhaust. With a directional control valve in a circuit, a cylinder's piston rod can be extended or retracted and work performed. If we need to control how fast the work is performed, we must control the flow rate.

1.6. A SIMPLE PNEUMATIC SYSTEM

When all of the pneumatic components which have been described up to this point are put together in proper order, they become the active elements of a simple pneumatic system (Fig. 1.13). The sys-

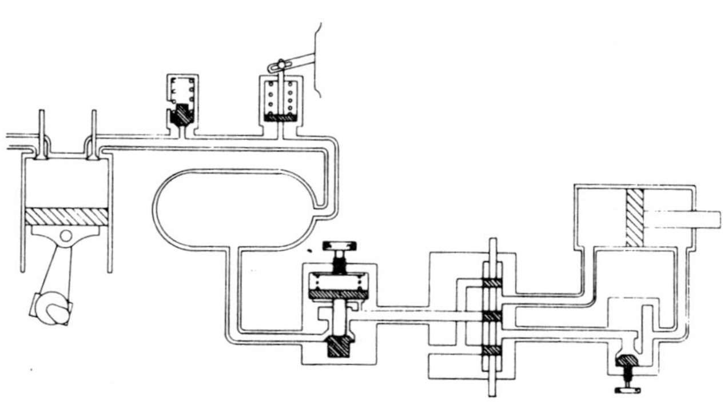

Figure 1.13

tem can perform useful work because pneumatic working energy in the system can be controlled. All power pneumatic systems operate on the principles that have been discussed up to now. In the remaining material dealing with pneumatics, we will concentrate on some of the typical components which are readily available for industrial pneumatic systems. Before we proceed, a few words about pneumatic symbols are needed.

1.7. PNEUMATIC SYMBOLS

The pneumatic components and elementary systems which have been shown to this point have been illustrated in a pictorial manner. System diagrams have been shown with simplified component cutaways to illustrate integral component operation. This technique is beneficial from an instructional viewpoint, but it is impossible from a workaday standpoint. In pneumatics, just as in other

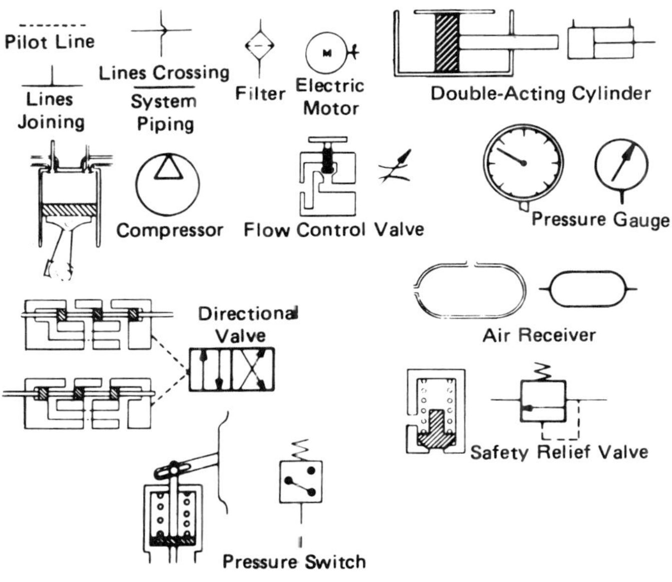

Figure 1.14

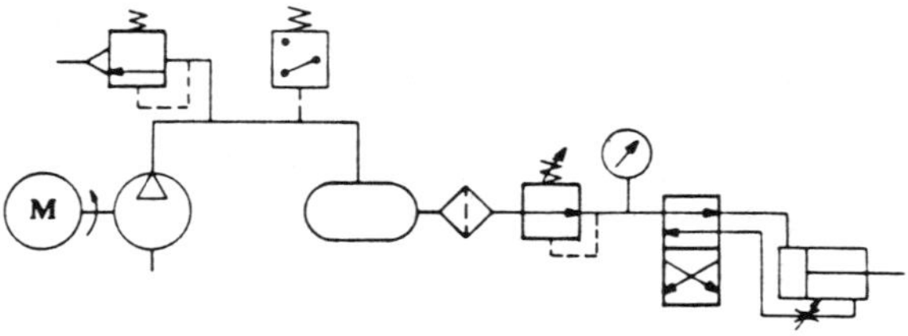

Figure 1.15

technologies, standardized symbols are used to "describe" components and systems. In this volume we use symbols from ISO 1219.1976 (E/F) and ANSI Standard Y32.10-1967(R1979) (Fig. 1.14).

The symbols for the components which have been discussed and the simple system which has been developed are illustrated by using graphic symbols for fluid power in Figure 1.15.

> *Note:* In addition to the components which have been discussed, the system which would normally be used consists of many others, such as air line lubricators, mufflers, and oil reclassifiers. Pneumatic systems are generally equipped with a prime mover such as an electric motor. Also, in order to achieve a degree of reliability, pneumatic systems should be protected from dirt with an air filter, entrained water should be removed, and lubrication should be provided where required.

The remaining chapters of this text will concentrate on one component in this system: the pneumatic directional control valve. The various types of valves, thin construction, sizing, troubleshooting, installation, and maintenance, and their application in many types of pneumatic circuits will be examined and illustrated in our pneumatic system. The other components in our pneumatic systems will be discussed in other volumes in this series.

2

Valve Types and Functions

2.1. INTRODUCTION

A directional control valve consists of a body with ports which are connected to internal flow passages by one or more movable parts that control the direction of air flow. These valves may also function to prevent or allow air flow. Other common applications of valves in circuits are to control the rate of air flow under a given set of pressure conditions.

2.2. FUNCTIONAL TYPES OF VALVES

One method of classifying a directional control valve is by the flow paths that are set up under its various operating conditions. Important factors to be considered are the number of external ports, the number of flow paths which are provided internally, and the interconnection of ports resulting from the different positions of the movable part. Let us begin our directional valve examination with the two-way valve.

2.2.1. Two-Way Directional Valve

A two-way directional valve consists of two ports connected to each other with passages which are open or closed (Fig. 2.1). In one extreme spool position, a passage from port A is open to port B; the flow path through the valve is open. In the other extreme, the large diameter of the spool closes the path between A and B; the flow path is blocked. A two-way directional valve has an on-off function similar to a water faucet's action. This function may be used to serve as an interlock or to isolate and connect various portions of the circuit. A very common construction of this valve, especially if the valve is electrically operated, is shown in Figure 2.1. Note that the valve can be normally open or normally closed, depending on the construction of the device.

2.2.2. Three-Way Directional Valve

A three-way directional valve consists of three ports connected through passages within a valve body (Fig. 2.2). They are shown here as port A, port B, and port C. If port A is connected to an actuator, port B is connected to a source of pressure, and port C is open to exhaust, the valve will allow the flow of air to, or exhaust from, a cylinder (or other device) connected to port A.

The function of this valve as shown is to allow air flow (pressurize) or block the air supply and allow air to exhaust from the actuator port. When the spool of a three-way valve is in one extreme position, the pressure passage is open to the actuator passage and air flows into it until that port is at the pressure level

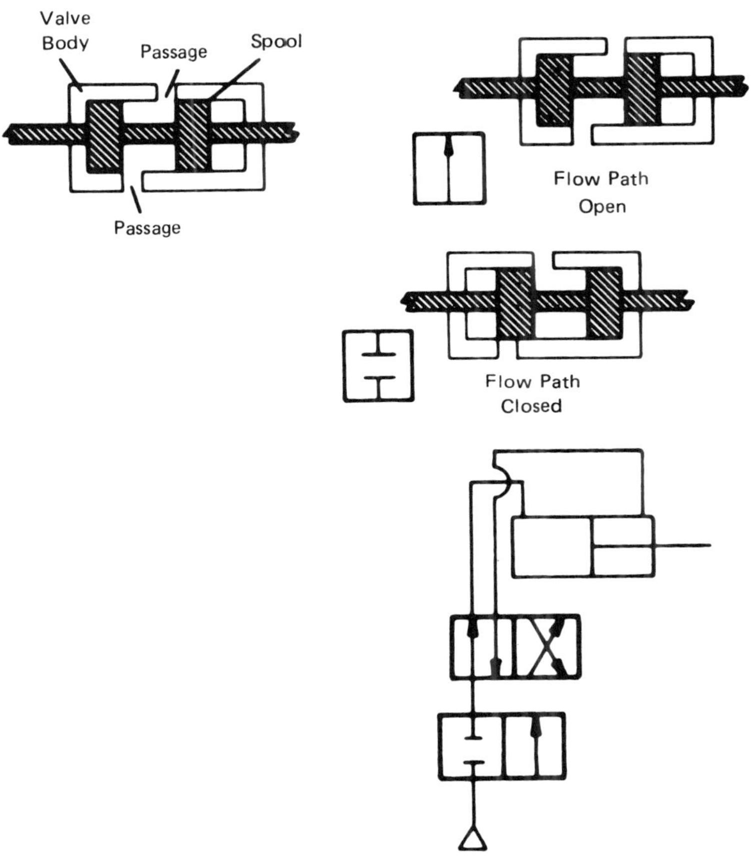

Figure 2.1 Control valves.

of the supply. When in the other extreme position, the spool blocks the inlet air passage, preventing flow, and the actuator port is opened to the exhaust passage. The valve can also be used to perform other circuit functions; for instance, it can be used to "select" the supply to a device connected to one of two pressures connected to ports B and C, or as a selector to direct flow to one of two devices connected to ports B and C.

These three-way valves may be used singly to control single-acting cylinders or in pairs to control double-acting cylinders. In

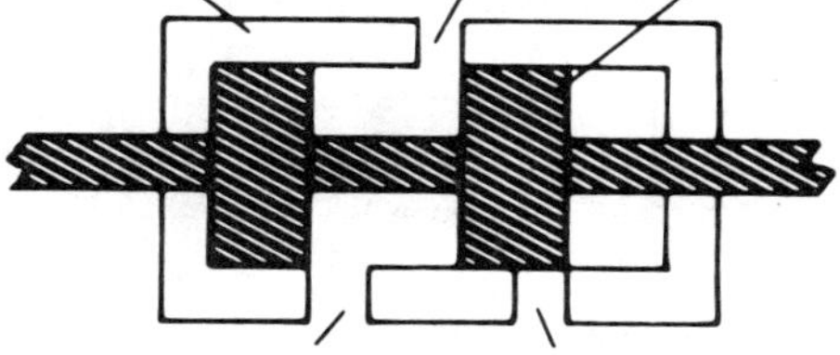

Figure 2.2

single-acting applications, a three-way valve often directs pressurized air to the cap end side of the cylinder (Fig. 2.3). When the spool is shifted to the other extreme position, flow and pressure to the actuator are blocked. But at the same time, the actuator passage is connected to the exhaust passage. The actuator must be returned by some other method. Here, in one case external force returns the ram. In the other case, an internal spring provides the return force. Also, three-way valves may be used in pairs to operate a double-acting cylinder, thus replacing a four-way valve. Paired three-way valves may be used instead of one four-way valve for one or more of the following circuit requirements.

1. High cylinder speeds are necessary. This requires air at high inlet pressures and an unobstructed low-pressure exhaust. Close coupling of three-way valves to the cylinder ports cuts down on exhaust back pressure and on pressure drop in lines, allowing greater cylinder velocities.
2. It is desired to save compressed air on high cyclic applications. The use of three-way valves, closely coupled to cylinders, eliminates the need to pressurize and exhaust pipe runs which do not contribute to the work done by the cylinder. This becomes important when large pipe sizes (½ inch and up) are used as higher cycle frequencies are required.

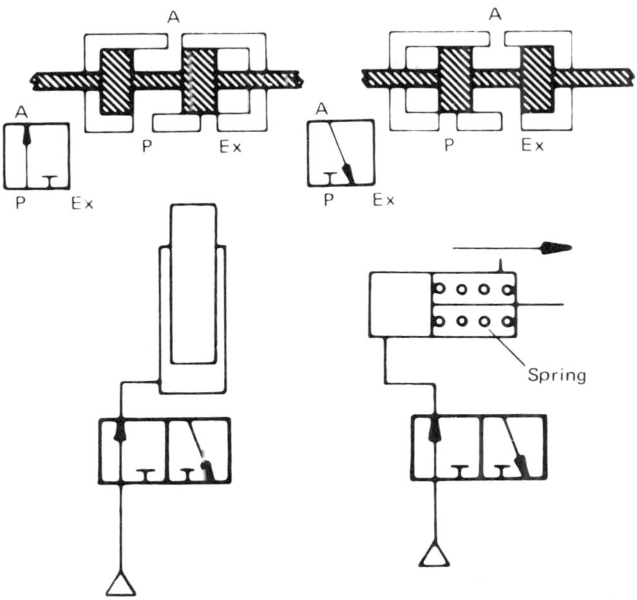

Figure 2.3

3. Intermediate positions are required. Here two valves are used with their normal position being the pressure port blocked or with both cylinder ports open to exhaust, giving a "float" type of cylinder control, when neither of the three-way valves is actuated. If only the valve at the cap end of the cylinder is actuated, the cylinder will extend; if only the valve at the rod end is actuated, the cylinder will retract. If both valves are actuated the cylinder will have pressure applied to both ends of the piston. Depending upon the pressure level supplied by each of the valves and the load characteristics, the piston may coast to a stop in some midstroke position or move to one end of its stroke. Other factors being equal, the most effective control valve-actuator installation uses valves as closely coupled to the actuator as possible. In some designs, the control valve is literally built into the actuator.

2.2.3. Four-Way Directional Valve

Perhaps the most common directional valve in simple pneumatic systems uses a pressure port, two actuator ports, and one or more exhaust ports. These valves are known as four-way valves since they have four distinct flow paths or "ways" within the valve body (Fig. 2.4).

A common application of a four-ported, four-way directional valve is to cause reversible motion of a cylinder or motor. To perform this function, the movable portion connects the pressure port with one actuator port. At the same time, the movable member connects the other actuator port with the exhaust port. This is a four-ported, four-way valve.

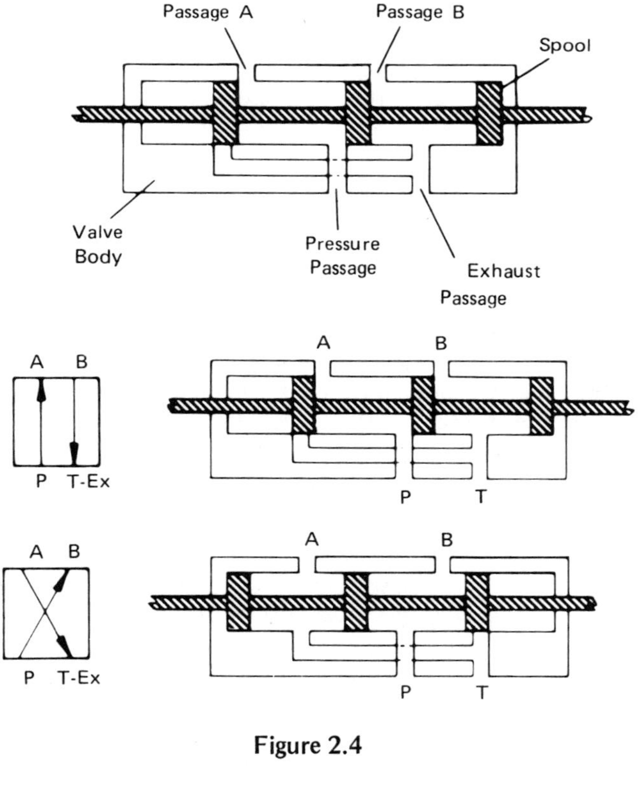

Figure 2.4

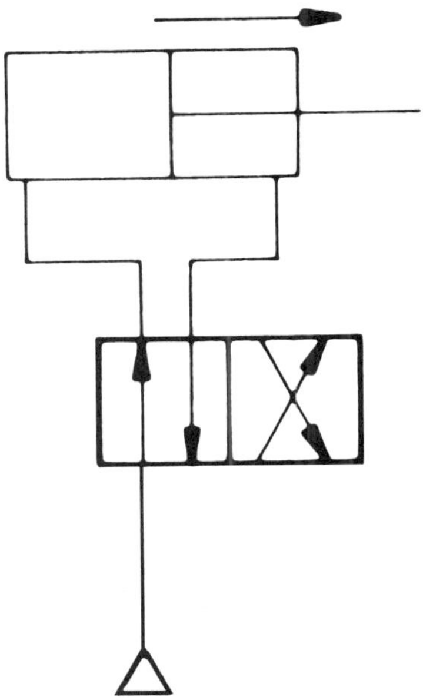

Figure 2.5

Four-way valves are also available with five external ports. Such valves provide the same basic control of flow paths as the four-ported version. In the fluid power field this is referred to as a "five-ported, four-way valve". This type of valve brings all flow paths to individual external ports.

The most common use of four-way valves is to control the motion of double-acting cylinders (Fig. 2.5). Four-way valves may also be used to control a pair of single-acting cylinders or to provide "flip-flop" control of air flow.

Alternatively, five-ported, four-way valves may be used to provide dual pressure control (Fig. 2.6). For example, one pressure can be applied to a cylinder when it is extending, and a different pressure when it is retracting.

Another way to apply a five-ported, four-way valve is with a common inlet pressure source and two exhaust ports. This allows individual control of the exhaust from each of the valve's outlet

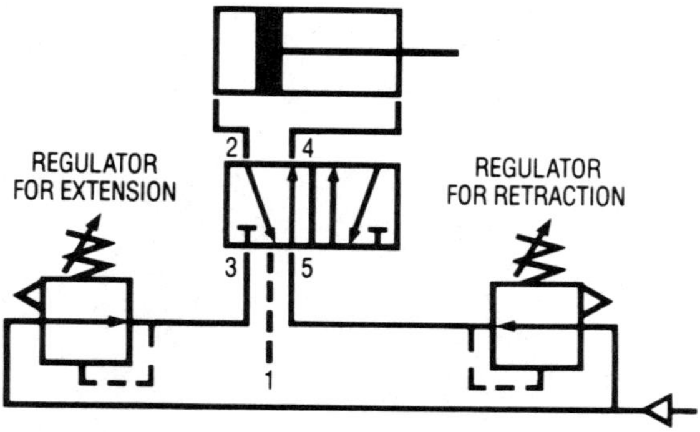

Figure 2.6 Dual pressure control.

ports. This method is often used to give a type of inexpensive "speed control" of pneumatic cylinders by appropriate restriction of the individual exhaust ports.

> Note that the "speed" of the actuator can be controlled only as long as all external loads, all friction forces, and the supply pressure are held constant. If any of these factors change, the actuator may become violently uncontrolled in its speed. Where this may be dangerous to the operator, the machine, or the process being controlled, other means of control should be used.

2.2.4. Spring Offset

The valves that we have been discussing could be of the two- or three-position type. A two-position directional valve generally uses an actuator to shift a directional valve spool to an extreme position. The actuator is then returned to its normal position by means of a spring. Two-position valves of this nature are known as spring-returned or spring offset valves in pneumatic systems.

2.2.5. Normally Open and Normally Closed Valves

Spring-returned two-way valves can be either normally open or normally closed (Fig. 2.7)—that is, when the actuator is not energized. In the normally open type fluid flow may pass and in the normally closed type fluid flow may not pass through the valve.

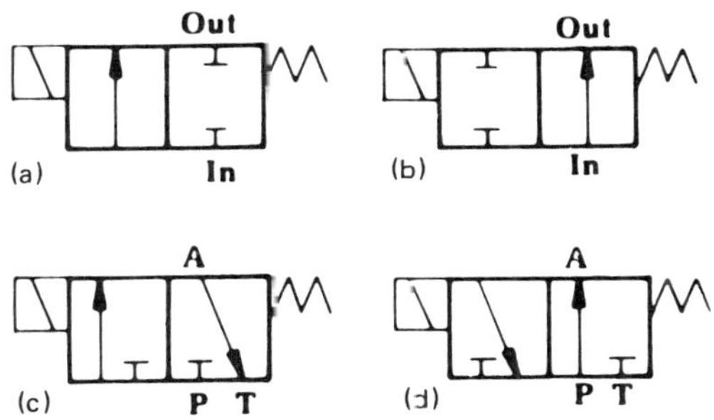

Figure 2.7 Spring-returned, solenoid-operated, two-way valves: (a) normally closed; (b) normally open. Spring-returned, solenoid-operated, three-way valves: (c) normally closed; (d) normally open.

In a two-position, three-way valve, since there is always a passage open through the valve, normally closed usually indicates that the pressure passage is blocked when the valve actuator is not energized.

Holding these valves in the actuated position without an external actuating force may cause a control circuit problem. If this is the case, then the use of a detented valve (without a spring return) should be considered. This will require the addition of another actuator to return the valve even over the detent resistance.

2.2.6. Detents

If either of two momentary actuators is used to shift the spool of a two-position valve, detents are sometimes used as a holding device to help keep the movable member(s) in the desired shifted position. However, conventional detents may not be sufficient to hold the movable member against impact, heavy vibration, or a sudden surge of air.

In our simplified illustration of a detent (Fig. 2.8), the spool is equipped with notches or grooves. Each notch is a receptacle for a spring-loaded movable part. In the detent illustrated, the movable part is a ball. With the ball in the notch the spool is held in position. When the spool is shifted, the ball is forced out of one notch and drops into another notch as the spool moves from one detented position to the other.

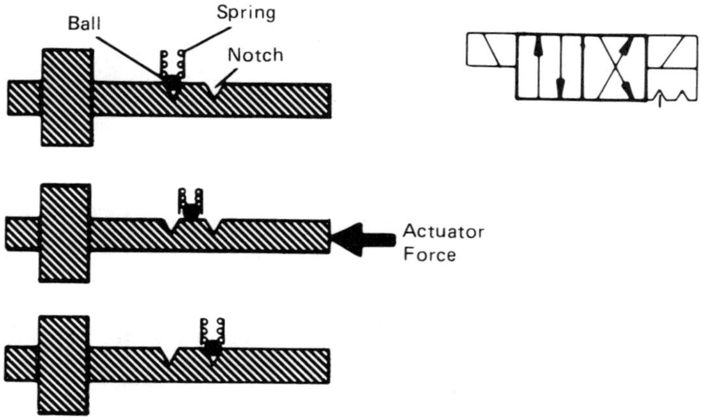

Figure 2.8 Two-position, solenoid-operated, detented, four-way directional valve.

Directional valves equipped with detents are found more frequently in pneumatic systems than in hydraulic systems.

2.2.7. Three-Position Valves

All the foregoing valve types are two-position devices providing two alternate flow paths—one in the normal position and the other in the actuated position. A three-position family of pneumatic four-way valves (in both four- and five-ported varieties) is commonly available.

In these valves, the two extreme positions of industrial four-way directional valves are directly related to the direction of movement (or force) of an actuator. They are the power positions of the valve. They control the movement (and direction of force) of an actuator first in one direction and then in the other direction. The third (center) position of a three-position directional valve is designed to satisfy a system requirement when no actuating signal is being applied. For this reason, a directional valve's center position is commonly referred to as a center or "neutral" condition.

There are several center conditions commonly available in three-position, four-way directional valves (Fig. 2.9). Some of the

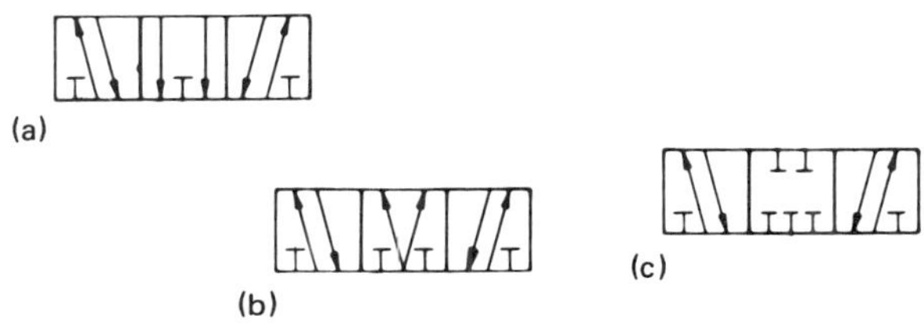

Figure 2.9 Center conditions available in three-position, four-way directional values: (a) exhaust center; (b) pressure center; (c) blocked center.

more common center conditions are: (1) both cylinder ports open to pressure (pressure center), (2) both cylinder ports open to exhaust with pressure blocked exhaust center, and (3) all ports blocked (blocked center). Let us examine each more closely.

The pressure center condition is illustrated in Figure 2.10. This type of valve may be used to alternately control the motion of two single-acting cylinders. When the valve is in the normal or neutral position, pressure is admitted to both cylinders, causing extension. If the valve is shifted to either extreme position, one cylinder will remain extended. When the valve is shifted in the opposite direction, the motion and force of the cylinders are reversed.

Next we will look at the exhaust center (Fig. 2.11). With this center condition, the cylinder is said to be free to "float." In other words, with the valve in its neutral state, both ports of the cylinder are opened to exhaust and the cylinder rod can be moved (subject to internal friction and external loading). When the valve is shifted to either of its extreme positions the cylinder extends or retracts as discussed before.

Last, the blocked center is examined (Fig. 2.12). This center condition blocks all working ports and is often called a closed or blocked center. Depending on the circuit design and actuator loading conditions, this center condition may provide a holding action on the device to which the valve is attached. With suitable controls and the compressibility of the air taken into account, this type of valve can be used to stop a cylinder along its stroke as it travels in either direction. Note that the actuator will move if the load changes or if there is even a small leak in the system.

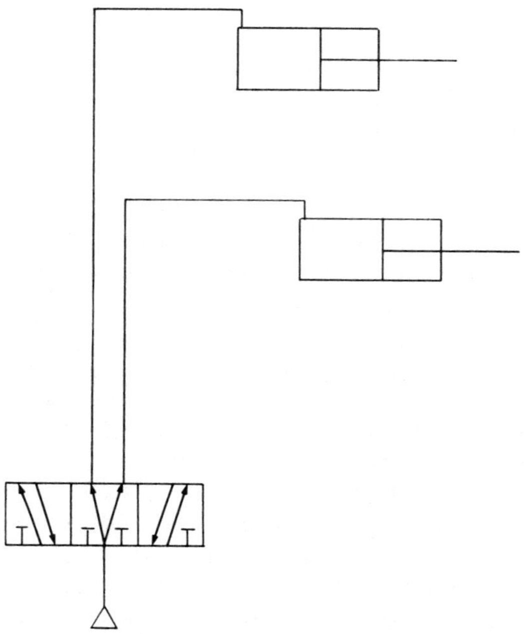

Figure 2.10

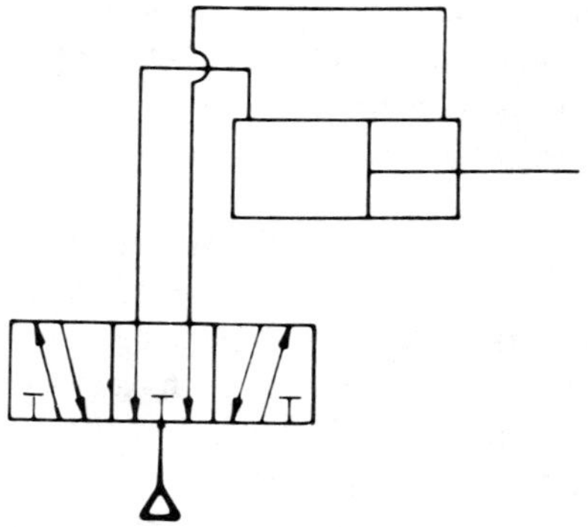

Figure 2.11

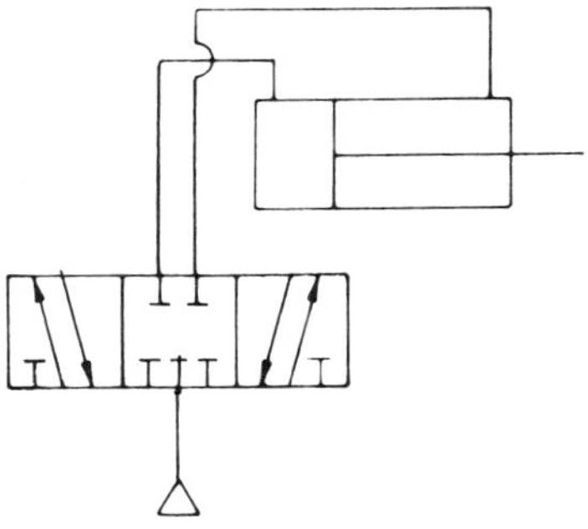

Figure 2.12

2.3. ADDITIONAL VALVE TYPES

Up to this point we have examined the pneumatic components known as air valves or directional control valves. Their function is to direct compressed air to a leg of a circuit or to an actuator. However, there are other valve types that can and are utilized to perform the same functions as the air valves. Their main or common application may not be that of a pneumatic directional control valve, but they will function very well in that capacity within their design limits. Two such valves are the pneumatic pressure regulator and the quick exhaust valve. Let us now review each of these valves.

2.3.1. Pressure Regulator

The primary function of the pressure regulator is to control the pressure supplied to actuators or other pneumatic devices and to prevent fluctuation caused by the compressor cycling between its high and low pressure settings. Ideally, the controlled pressure must be set sufficiently below the lowest expected pressure available to allow for a 3 to 5 psi control pressure differential at the time the air supply pressure to the regulator is at its minimum and the air flow through the regulator is at its maximum.

Depending on the application, a pressure regulator may have to meet the requirements of:

1. Maintaining output (secondary) air at a constant pressure regardless of pressure variation in the upstream pressure and independent of flow
2. Reducing the pressure to a level which meets the work requirement and thus minimizes the amount of wasted air

It should be noted from the regulator symbol (Fig. 2.13) that this pressure control is normally open, and secondary pressure is sensed for control purposes.

Acting like a flow control valve, a pressure regulator positioned after an air receiver tank allows compressed air to expand downstream (Fig. 2.14). As pressure downstream rises, it is sensed at the control surface of the regulator through an internal passage at the underside of a piston or diaphragm. This control surface has a large area exposed to downstream pressure and is therefore sensitive to pressure fluctuations. When downstream pressure is high enough to overcome the control force, the control element moves upward, causing the valve to close. The valve throttles off flow to the downstream passage and does not allow pressure to build beyond the set (control) pressure. In this way, a relatively constant pressure is made available to an actuator connected downstream. Note that the adjustable spring acting against the piston exerts a different force for any variation of the adjustment device and thus allows for variations in the controlled pressure. Often overlooked is the fact that at a higher flow rate, the spring is extended and thus presents a different control force on the poppet. This is referred to as a regulator's spring rate. (See Sect. 2.3.2.)

A venting-type regulator (Fig. 2.15) limits downstream air pressure to a level lower than the supply pressure at the regulator's inlet in a manner similar to that described above for a nonventing regulator. It has an additional function in that it acts as a small relief valve for the control portion of the circuit in case of any pressure buildup.

Note: The venting relief flow capability is generally inadequate for critical operations where large flow rates in the venting mode will be required. It is suggested that a full flow relief be used whenever large volumes need to be released quickly.

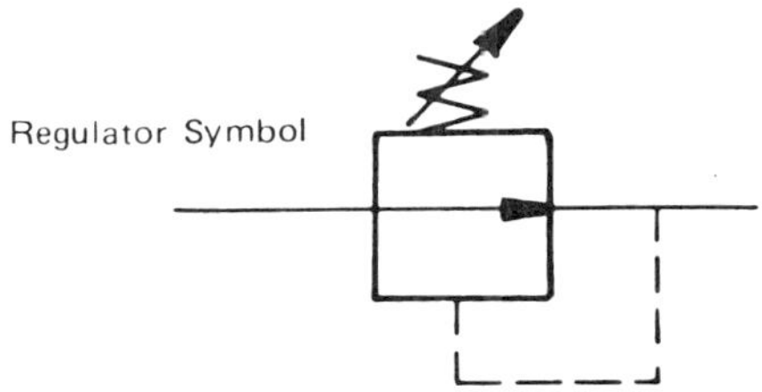

Figure 2.13

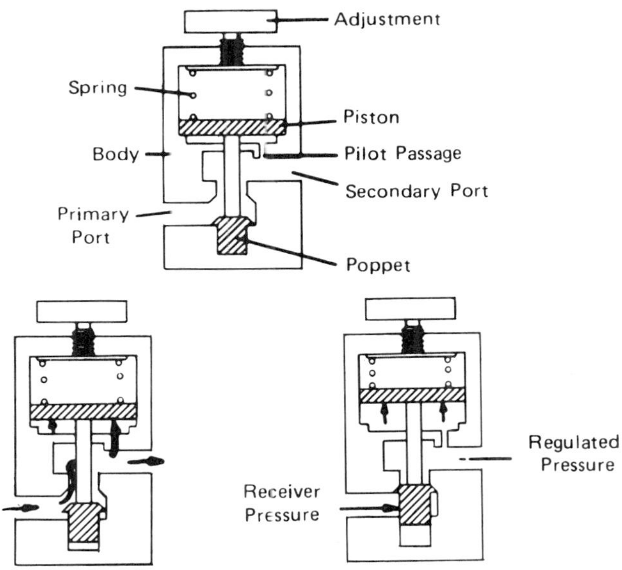

Figure 2.14

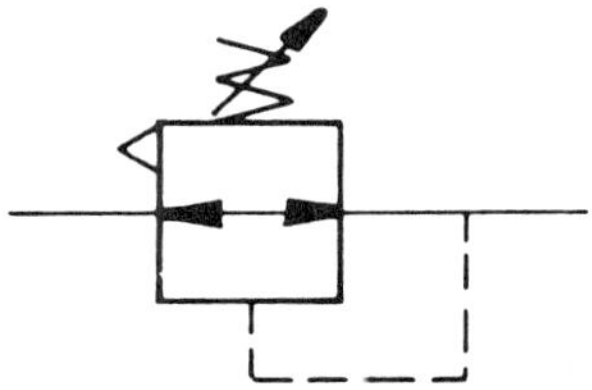

Figure 2.15 Venting-type regulator symbol.

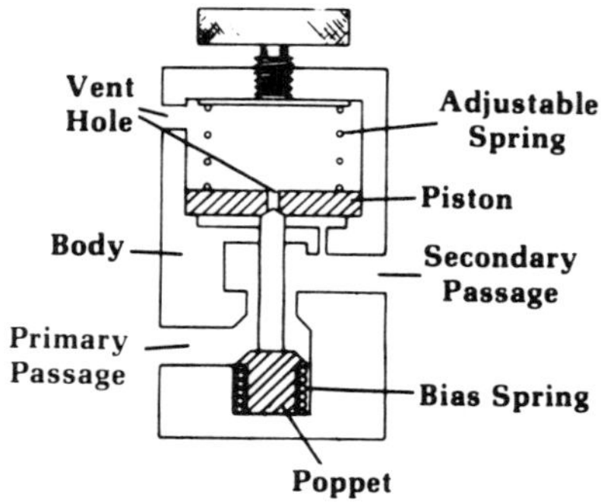

Figure 2.16 Venting-type regulator.

A typical venting-type regulator (Fig. 2.16) basically consists of a body with primary and secondary passages, a poppet with light bias spring, a pressure-sensing device with a vent hole, and an adjustable control force, usually a spring. A regulator of this type should have as little restriction to flow as possible in the venting flow direction. A large vent area and passages will create a minimum pressure drop across the regulator in the venting mode. A circuit that requires venting for safety and incorporates this type of regulator could function without a relief valve. However, this is not recommended, because the limited vent flow capacity may create an unsafe condition for the operator and/or the machine. No system should be operated in such an unsafe condition. There should be no substitute for a properly sized and adjusted relief valve. Because of this, savings in labor and materials due to fewer components and connections would be realized.

Venting-type regulators will maintain a set secondary pressure. If for some reason the secondary pressure exceeds a valve setting, a pressure-sensing device will open a vent hole, reducing the excess secondary pressure to the correct value.

It should be pointed out that venting, as defined here, is not the same as reverse flow. Reverse flow regulators, unlike many venting regulators, have the capability of passing flow from the normal outlet (controlled port) to the inlet (pressure) port. This is passing all flow, not just enough to maintain a set pressure. Such a regulator may be a very efficient pressure control device when connected to an actuator port as it will control pressure at the closest possible point of the circuit. At the actuator port, all pressure losses ahead of the regulator during flow are then a part of the supply pressure variation and do not appear as losses between the regulator and the actuator. Similar circuit enhancements can be devised by use of a regulator which has a remote downstream sensing capability.

A venting-type regulator controls downstream pressure in the same manner as an ordinary regulator while air is being supplied to the system. As pressure rises downstream of the regulator, it is sensed on the underside of the piston or diaphragm (Fig. 2.17). When the force at the piston or diaphragm is large enough to move against the force of the adjustable spring, the piston and poppet move upward, throttling or closing the valve and limiting maximum downstream pressure. The difference in operation between a nonventing and a venting regulator is the venting action taken when a secondary pressure rises above the setting of the regulator. This higher pressure could be the result of a pressure buildup caused by leakage (either across the regulator seat or internal to the actuator), thermal expansion of air, or a load pushing externally against a cylinder piston. It is most often observed as the result of resetting the regulator to a lower downstream pressure setting.

When a higher-than-set pressure appears downstream in a venting-type regulator, pressure acts on the piston or diaphragm, pushing it up (Fig. 2.18). At the same time, the poppet is pushed up by

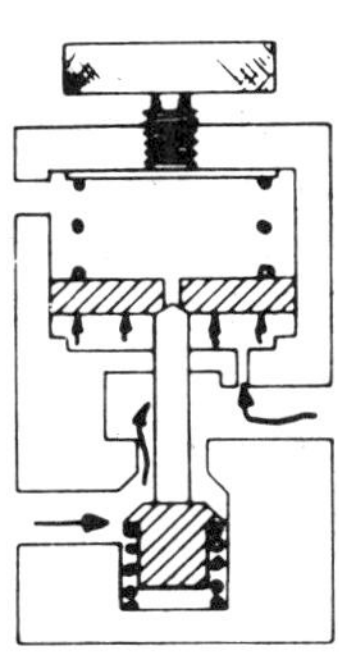

Figure 2.17 Regulator regulating.

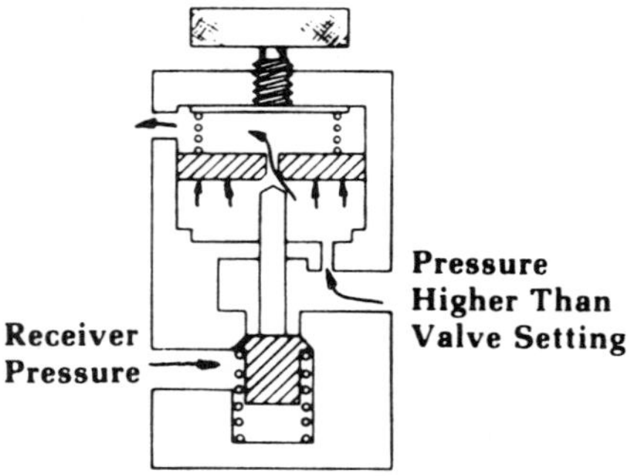

Figure 2.18 Regulator venting.

its light bias spring. With the poppet seated, the piston or diaphragm continues to move until a vent hole is opened. Excess pressure is thus bled off through this hole to the atmosphere.

When downstream pressure returns to the desired level, the piston or diaphragm reseats onto the poppet stem and pressure regulation resumes. Most pressure regulators found in industrial pneumatic systems are of the venting type for the purpose of easy adjustment.

As indicated previously, there is another very common type of pressure regulator. This regulator uses a diaphragm regulator in place of a piston (Fig. 2.19). Operation of a diaphragm regulator is similar to that of a piston type. An example of a venting-type diaphragm regulator is illustrated. Next we should examine a device called a pilot-controlled regulator.

A pilot-controlled regulator replaces the adjustable control spring for the principal air flow with air pilot pressure. One advantage of a pilot-controlled regulator over a standard regulator is that secondary pressure can be remotely adjusted. Another is that the control force on the control spring side of the piston or diaphragm is almost constant, regardless of the valve position. Because the "spring rate" of the controlling air pressure is almost constant, a pilot-controlled regulator has a distinct control advantage over a spring-loaded type, which has a "steep" spring rate depending on valve opening, resulting in a variation of control pressures dependent on flow rates.

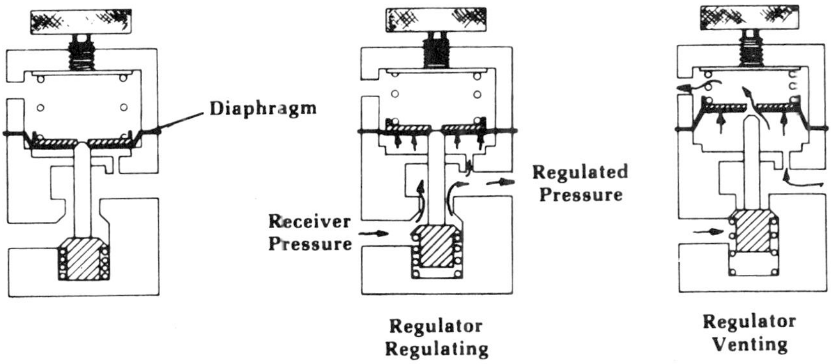

Figure 2.19

A pilot-controlled regulator (Fig. 2.20) basically consists of a valve body with primary and secondary passages, a piston or diaphragm, and a poppet biased closed by a light spring.

A pilot-controlled regulator operates in the same basic manner as an ordinary regulator. The difference is that pilot air pressure acts as the control force against the piston or diaphragm instead of a spring.

As pressure in the downstream circuit of the regulator rises, it is sensed internally at the underside of the piston or diaphragm

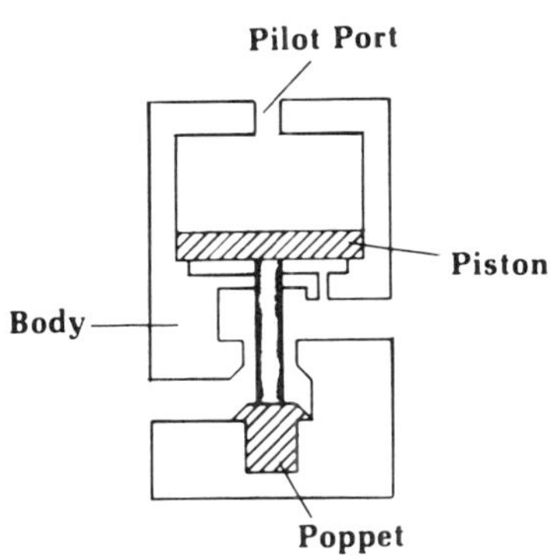

Figure 2.20 Pilot-controlled regulator.

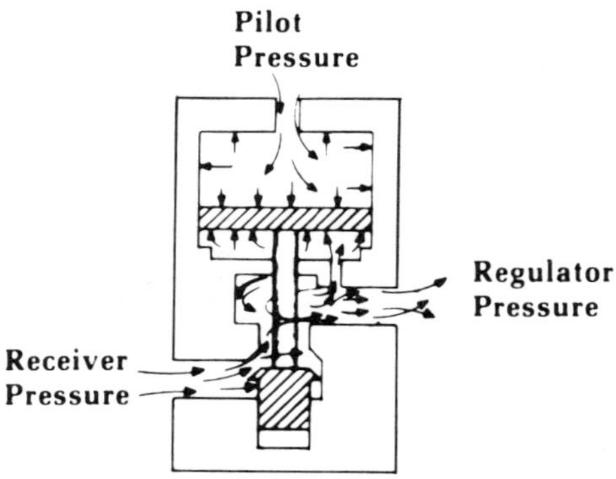

Figure 2.21

(Fig. 2.21). When the force generated at the bottom of the piston is greater than the force developed on top of the piston from the pilot pressure, the poppet and piston move toward the poppet seat, regulating flow and thus pressure. Pilot-controlled regulators, of the piston or diaphragm type, can be venting or nonventing and may be designed for reverse flow (Fig. 2.22). To achieve proper pressure regulation of a circuit, we must select the proper type of regulator and size it correctly.

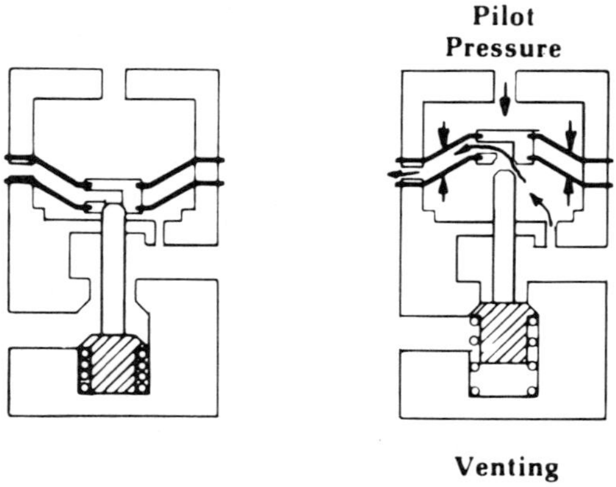

Figure 2.22 Diaphragm pilot-controlled regulator—venting type.

2.3.2. Pilot-Controlled Regulators as Directional Control Valves

By alternately applying and removing pilot pressure, a pilot-controlled regulator can function as a two-way (on-off) directional control valve. It can be operated to respond to a series of preselected pilot pressures. Also, a venting-type regulator could be used as a three-way valve. However, caution must be exercised when attempting this. The vent passage must be large enough to pass the required reverse flow rate to the atmosphere. In this application the regulator would freely pass flow from the primary port to the secondary port. When secondary pressure reaches the valve setting the valve would shift to a closed position. When the secondary pressure began to exceed the valve setting, the large vent hole would be uncovered. All flow from the secondary would pass to the atmosphere. The regulator would be performing the function of a two-position, three-way, normally open air valve. Many interesting circuit possibilities can be devised by the resourceful circuit designer.

2.3.3. Quick Exhaust Valve

Several special-purpose valves are used in pneumatic circuitry. A quick exhaust valve (Fig. 2.23) is an example. It is most commonly used in conjunction with a three- or four-way directional control valve to increase the exhaust flow (and thus generally increase the rod speed of the cylinder).

The quick exhaust valve typically consists of a resilient disk in a body. The resilient disk "shuttles" from side to side, changing the free flow paths through ports in the body.

The quick exhaust valve is typically used in conjunction with three- or four-way directional control valves (Fig. 2.24). When its inlet port is connected to the supply pressure, it automatically seals its exhaust port and allows pressure to enter the cylinder. If the valve is of the "floppy disk" design, a control valve is shifted so that it can exhaust the supply to the cylinder. The resilient disk moves away from the exhaust port and allows the trapped air in the cylinder to quickly exhaust directly to the atmosphere. This eliminates the need for exhaust air from the cylinder to flow through long, often somewhat restrictive, lines and through a main control valve. Quick exhaust devices should be connected

Figure 2.23

(in juxtaposition) to the cylinder port to give high cylinder velocities.

One application for a quick exhaust valve is using a single-acting spring return cylinder. With a standard circuit, many times the return speed may be too slow. This may be due to inadequate spring force, high frictional resistance due to the load, restriction of exhaust flow, or a combination of several of these factors.

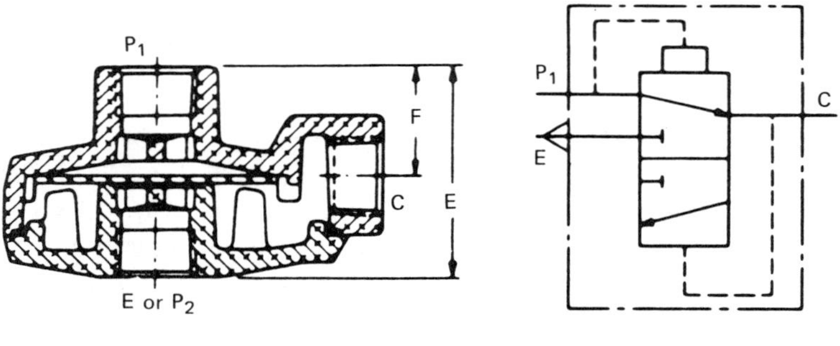

Figure 2.24 Quick exhaust valve.

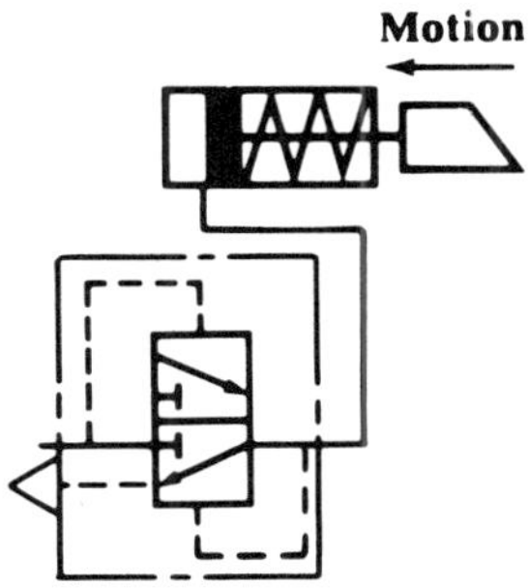

Figure 2.25 Retraction of single-acting cylinder.

Figure 2.25 shows a single-acting cylinder with a quick exhaust valve closely coupled to the cylinder. This permits greater cylinder speeds than would exist if the air from the cylinder were exhausted through the directional control valve, but without increasing its size. During extension of the cylinder, the quick exhaust is part of the supply system that feeds air into the cylinder (Fig. 2.26). However, during the return stroke, the quick exhaust shifts as soon as the pressure between it and the directional control valve drops below the pressure at port C. This permits free exhausting of the air from the cylinder directly to the atmosphere, bypassing the

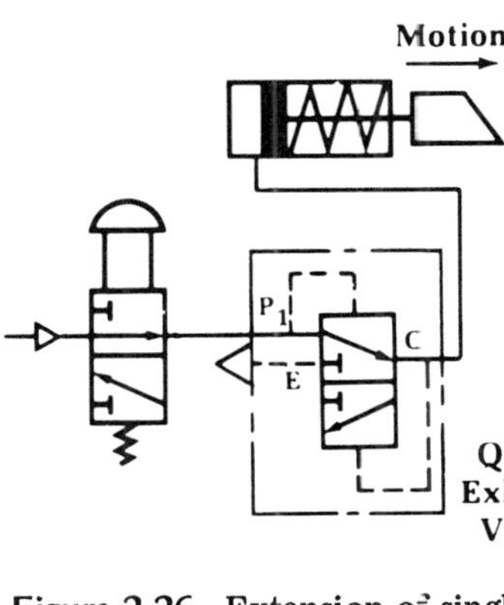

Figure 2.26 Extension of single-acting cylinder.

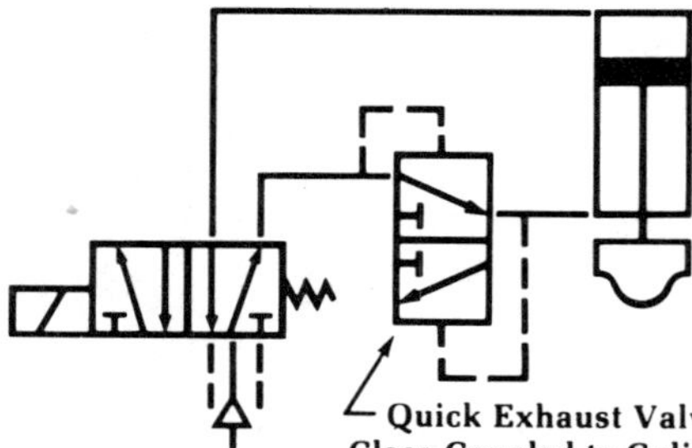

Figure 2.27 Quick exhaust valve added to double-acting cylinder for very rapid speed. The cap end of the cylinder must have a pressurized volume to insure that the cylinder will deliver an impact blow. There is no control over the rate of cylinder rod extension.

directional control valve and all the interconnecting lines. Quick exhaust valves may also be added for double-acting cylinder applications. An application that will deliver an impact blow is illustrated. In this circuit, a quick exhaust valve was incorporated at the cap end for high return speed (Fig. 2.27). Note that the volume of air under pressure in our example will act as an air spring.

Many quick exhaust valves may also be used as shuttle valves. Shuttle valves are automatic flow path selectors which allow the

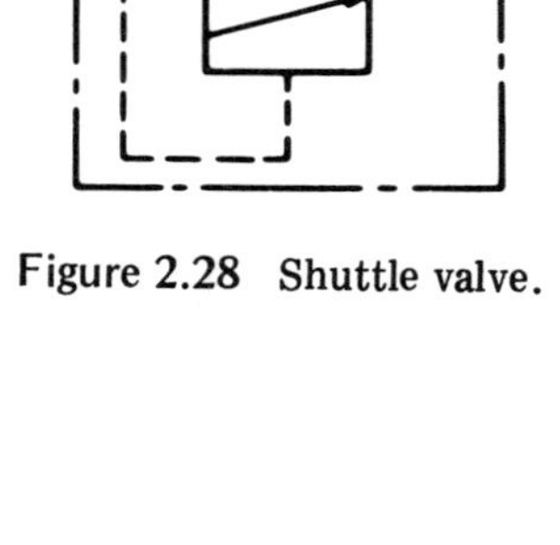

Figure 2.28 Shuttle valve.

higher of two pressures to be directed into a flow path. The use of such devices is common in control circuits to "select" the higher sources of pressure within the circuit (Fig. 2.28). However, when exhausting a large volume of air through a large valve, especially through a quick exhaust valve directly to the atmosphere, unwanted air noise may become a problem. When using a quick exhaust valve, in most applications, a muffler should be installed closely coupled to the valve's outlet port.

3

Valve Construction

3.1. INTRODUCTION

Basic valve design should be considered when selecting pneumatic directional control valves. Since they control direction, they must be adequately sealed for the application. The sealing function may be accomplished by shear action or seating action elements. The discussion following describes the typical characteristics of each of the valve types. We do not imply—nor should the reader infer—that any of the valve types discussed may or may not include the characteristics attributed as being typical to any particular design. Specific characteristics depend upon the valve manufacturer's designs. Special applications should be discussed with the valve manufacturer.

3.2. SHEAR ACTION VALVES

The shear action type of valve controls flow by means of an element that slides across the flow path. There are four basic types: (1) sliding plate or rotary disk, (2) packed spool, (3) packed bore, and (4) lapped spool.

3.2.1. Sliding Plate (Rotary Disk)

The first type, the sliding plate or rotary disk (Fig. 3.1), uses pressure unbalance to force the sealing mechanism (commonly called a D-slide) against a mating surface. The effect is to control the flow of air to and from desired ports and seal the flow off from others. These valves can provide two-, three-, or four-way action.

The sliding plate principle is available in a modified form called a rotary disk. This utilizes the shear sealing principle of the D-slide valve, but allows control by rotary motion as an actuating means. Desired porting connections are achieved by rotating the D-slide, which now may be termed a rotary disk. This can be seen in Fig. 3.2.

The following are typical characteristics of sliding plate valves.

1. Valve mechanisms "wear in" as they are used. Valves are capable of many millions of trouble-free cycles, even under adverse conditions.
2. Materials may be selected which are not affected by lubricants or contaminants present in compressed air lines.
3. Wear tends to keep mating parts in contact, thus controlling leakage over long periods of use.

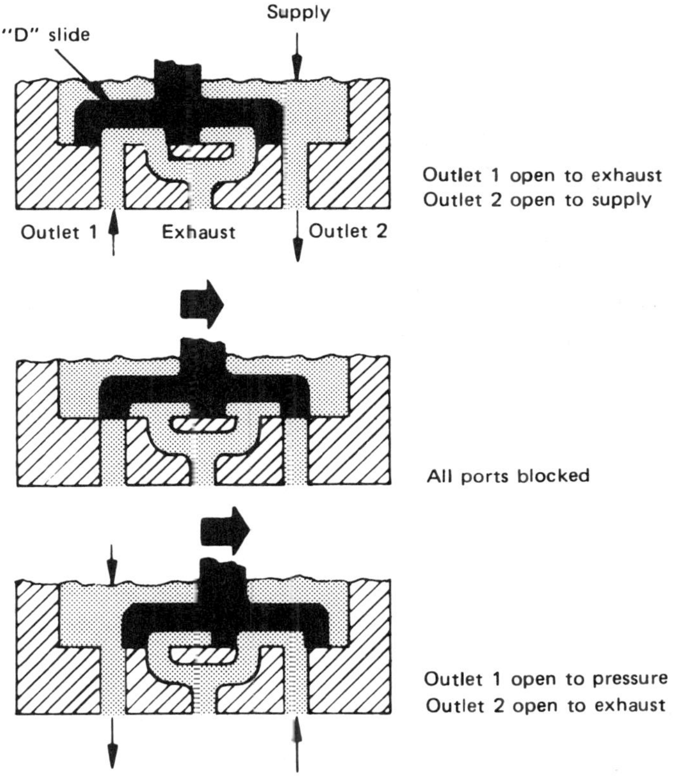

Figure 3.1 Typical "D" slide type valve.

4. The valve tends to exclude foreign material from lodging between the mating surfaces.
5. The valve is not seriously affected by temperature extremes.
6. The valve can be used for flow control and directional control.
7. This type of valve has good air flow throttling in proportion to the movement of the control valve.

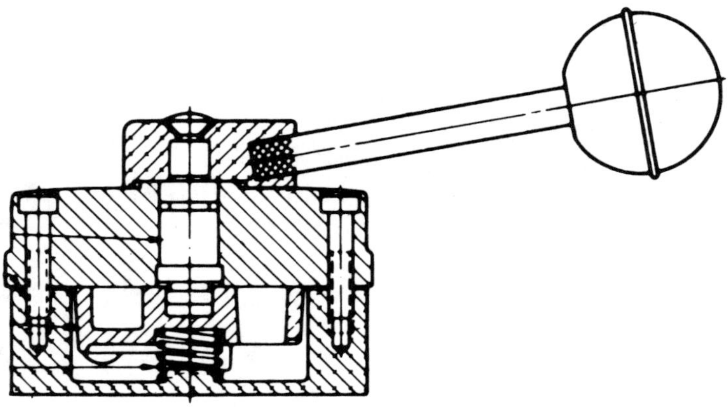

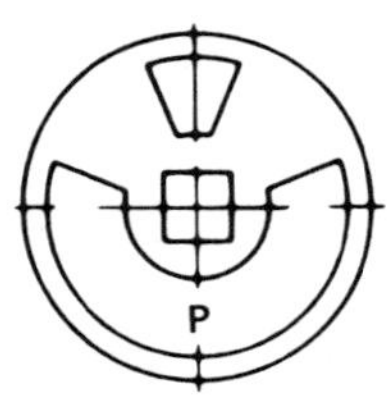

TYPE P DISC
3 POSITION 90 MOVEMENT
ALL PORTS BLOCKED IN NEUTRAL

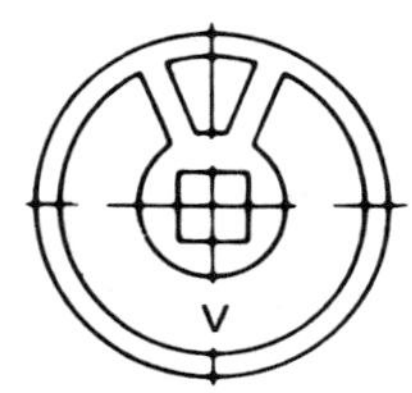

TYPE V DISC
3 POSITION 90 MOVEMENT
INLET BLOCKED CYLINDER PORTS OPEN
TO EXHAUST IN NEUTRAL

Figure 3.2

To minimize circuit problems with sliding plate valves, consider these factors:

Large forces may be required to shift the valve slide, especially at high pressure, when using designs with large pressure-sealed areas, which are not at least partially pressure balanced.

Units may require long actuator stroke lengths to obtain full flow capabilities (in some applications this may be a distinct advantage).

The valve must be continuously lubricated for maximum life. (However, present-day material technology allows the component designer to have lubrication built into the material.)

Design may not provide bubble-tight sealing.

The sealing member (D-slide or disk) can be forced from its mating surface if the pressure under the plate exceeds the line pressure. Some valve designs may not be suitable for positioning or stopping cylinders where such pressure conditions can occur.

3.2.2. Packed Spool

Another type of shear action valve is the packed spool valve. The spool of this type of valve is provided with seals to effect almost leakproof sealing. The packed spool valve design uses a resilient seal around the spool lands. The spool and its resilient seal works in a rigid bore (Fig. 3.3).

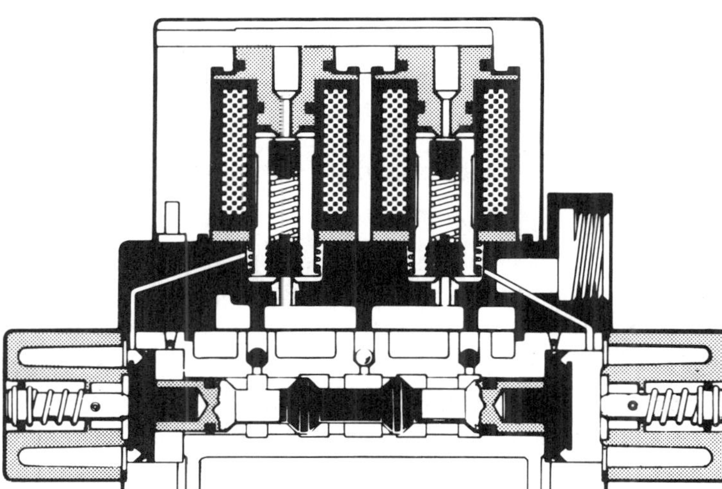

Figure 3.3

Packed spool valves typically offer the circuit designer the following characteristics:

1. Packed spool design permits maintenance of the seal shape and size, thus virtually providing a positive bubble-tight seal for many cycles.
2. Packed spool valves are less vulnerable to "varnishing" in place than are the lapped spool valves because only small areas are in contact or very close fit within the bore. Further, even a slight shifting force on the seals causes the elastomer to deflect and "peel" away from a bonding varnished area.
3. Packed spool valves are less affected by improper installation (improper torquing), where distortion of the body may cause changes in the contact between the bore and the spool.
4. Packed spool valves are less sensitive to contamination than are lapped spool valves.
5. Maintenance is generally less costly from the standpoint of cost of parts replacement.
6. The valve is usually of a balanced spool design. It tends to stay in place when shifted, which is an advantage if power is lost.

To minimize circuit problems with packed spool valves consider these factors:

Compatibility of seals and any fluids which may be carried by the air stream must be checked. Swollen seals require excessive shifting force on the spool. Shrunken seals will increase leakage.

Some designs may use seal materials which tend to adhere to the surface against which they seal. Seals working against copper bearing surfaces (e.g., brass, bronze) are more likely to "bond" after long set times than are seals working against aluminum or many types of plated surfaces.

Extreme temperatures may cause the seals to change size or become hard.

3.2.3. Packed Bore

Another very common directional valve type is the packed bore design. The packed bore valve has a spool with rigid sealing lands working in a bore that has several stationary elastomeric seals which provide sealing (and thus isolation) between ports (Fig. 3.4).

Packed bore valves typically offer the circuit designer the following characteristics:

1. The design is available in a variety of flow path patterns in most porting and actuating configurations.
2. The design results in a balanced spool. Shifting forces are generally only slightly higher than for metal-to-metal designs. Sudden pressure surges cannot cause the valve to lose its sealing capabilities.
3. The resilient seals make the valve less vulnerable to abrasion by foreign material than metal-to-metal designs.
4. Valve spools have limited areas in contact and thus less tendency to "varnish" in place than metal-to-metal designs.

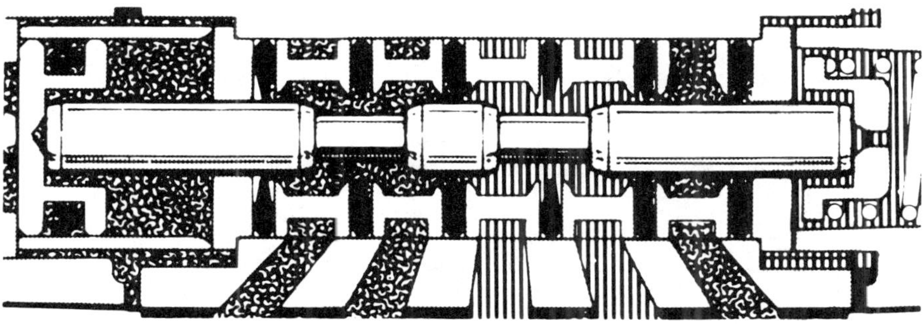

Figure 3.4 Packed bore.

5. For greater life, lubrication is suggested.
6. The spool will not bind under ambient temperature encountered in industrial application.
7. The seals may be changed without changing mating parts.

To minimize circuit problems, consider these factors for packed bore valves:

Compatibility of seals and airborne liquids must be checked. Swollen seals often require excessive shifting force on the spool. Shrinking seals may also result in excessive shifting force to move the spool. Shrinking seals will often introduce leakage.

Some designs use seal materials which tend to adhere to the surface against which they seal.

Extreme temperatures may cause the seals to change size or become hard.

These spool-type valves are provided with seals to effect virtually leakproof sealing. Packed spool valves are arranged with a relatively rigid spool with resilient seals working in a rigid bore. Packed bore valves have relatively rigid spools working in a rigid bore provided with resilient seals. In either design, the rigid parts may be metal, ceramic, plastic, or another relatively rigid and dimensionally stable material (Fig. 3.5).

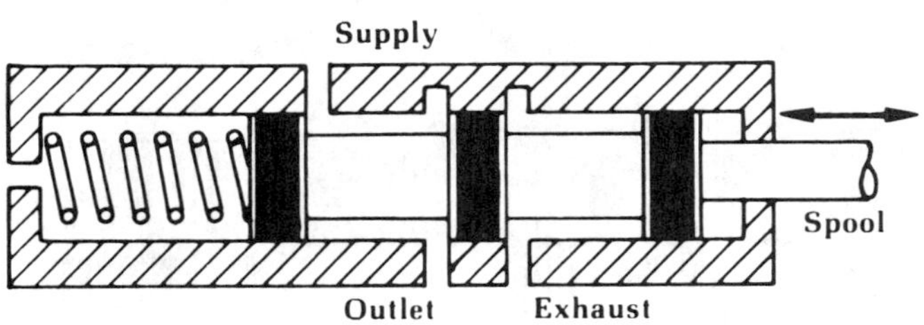

Figure 3.5 Typical packed spool type valve.

Selection of seals in either case is important. The elastomer must be resilient enough to provide sealing, yet tough enough to provide resistance to abrasion. It must be dimensionally stable under all operating temperatures and in the presence of various fluids which may be carried through the valve.

3.2.4. Lapped Spool Valves

The lapped spool valve is also classified as a shear action type. This design depends on a very close fit between mating parts to control the flow of air from one port to another. This design does not give bubble-tight sealing since the spool and bore must have at least a minimum clearance to enable them to move at a low friction level and tolerate some warpage of parts (Fig. 3.6).

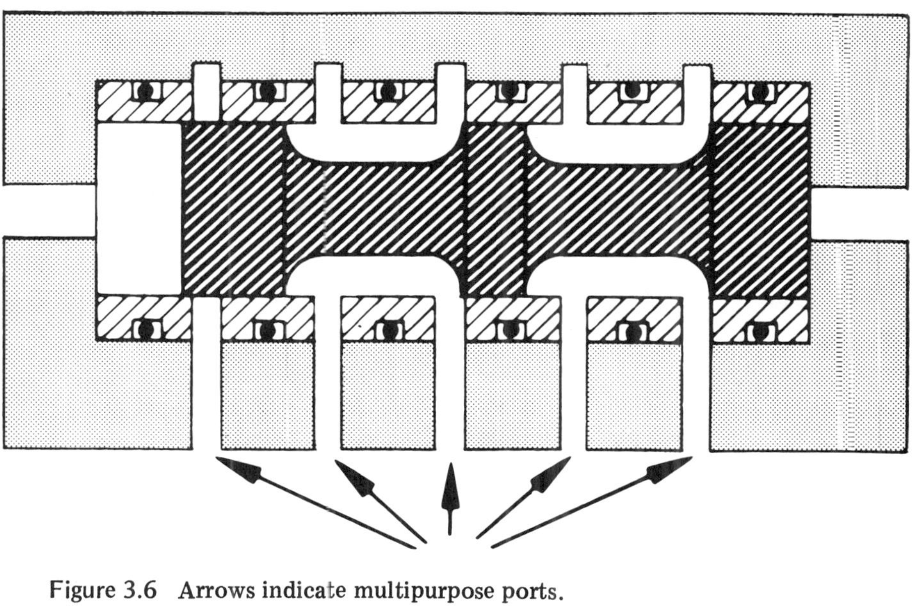

Figure 3.6 Arrows indicate multipurpose ports.

The following are characteristics of the lapped spool valve:

1. The design is capable of providing almost any flow path pattern desired in a variety of porting and actuating configurations.
2. External force normally required to shift a balanced spool is low. This is desirable when direct actuation of the spool is required.
3. The force required to position the spool tends to remain constant during a shifting stroke. This helps the spool to complete its stroke once "break-loose" friction has been overcome.
4. The design can easily be made to prevent interconnection of the pressure, outlet, and exhaust ports while the valve is being shifted from one flow path condition to another. The elimination of crossover flow while the spool is in transit reduces the probability of shift failure when operating at low pressures as a pilot-operated valve, and can eliminate spurious signal pulses.
5. Because the spool is balanced, sudden pressure surges resulting from external forces on cylinders cannot cause the valve to lose its sealing capabilities.

To minimize circuit problems with metal-to-metal spool-type valves, consider these factors:

Long stroke requirements of spools may entail excessive travel of mechanical actuators to complete the shifting of the valve.

Closely fitted parts are vulnerable to the entrance of foreign matter: particulate and gumlike materials are especially troublesome. This can cause rapid wear and leakage, or cause the parts to stick.

Units of this type generally require good lubrication and filtration. In some applications, the manufacturer may suggest that it is best to run on dry unlubricated air.

Oxidized airborne lubricant from a compressor or other material carried down an air line may cause the closely fitted parts to varnish in place. This occurs particularly when valves are allowed to remain idle for long periods of time, as over weekends. Such applications often require the use of special oil-removing filters.

Initial cost is high. Maintenance costs are higher than for other types of design.

Improper installation (improper torquing) may cause the valve spool to fail to shift, due to valve body distortion.

If the valve is subjected to vibration parallel to the axis of the main spool, detents may be needed.

3.3. SEATING ACTION VALVES

Up to this point, the discussion has been limited to shear action valves. We would now like to discuss poppet types (Fig. 3.7). Directional valves that employ poppets for their movable part(s) are suited for use where high flows are encountered. Since a poppet opens up a relatively large flow passage in a short stroke, poppet valves have an inherent characteristic of fast response with minimum wear. Poppet valves for pneumatic service usually employ resilient seals to provide tight sealing. Most poppet valves found in industrial applications are three-way. Usually a four-way poppet valve is made up of two three-ways working in a common body (Fig. 3.8).

Characteristics of a poppet valve include:

1. Poppet valves offer rapid cycling capabilities. Light moving members require only a short stroke to provide maximum flow opening. Cyclical frequencies to 3600/minute have been reported. Their use on high-speed welding and packaging equipment is evidence of this.
2. Short stroke of the valve gives minimum wear and maximum life capabilities.
3. Resilient seals give tight shutoff of the flow path and help absorb the kinetic energy of the moving members.

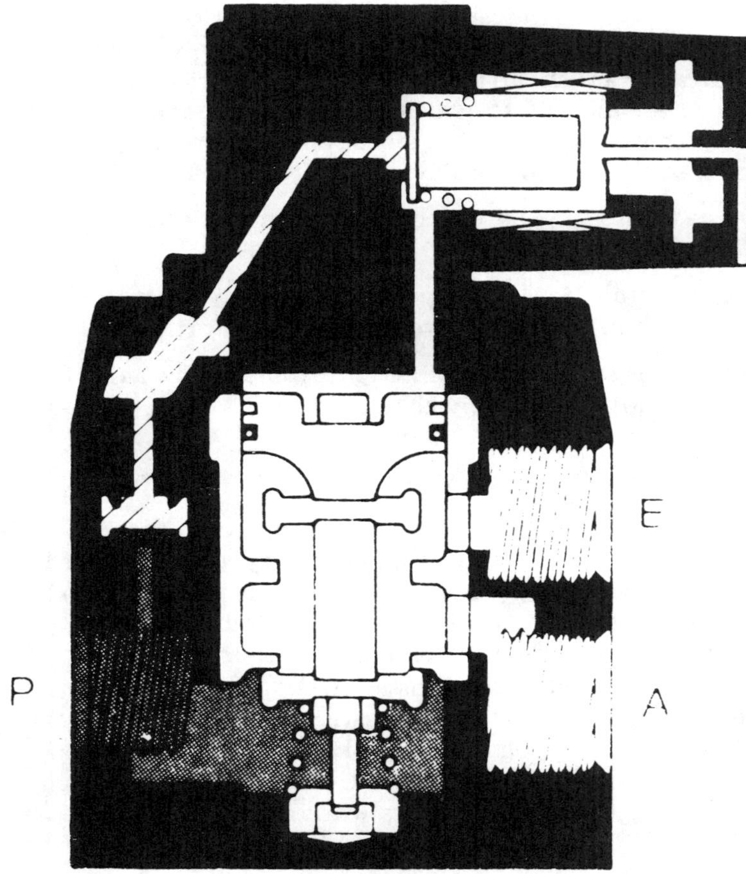

Figure 3.7 P, poppet valve; A, action valve.

4. The design is resistant to damage by foreign matter carried through the air line since the seats are self-cleaning.
5. Maintenance of the valve is readily accomplished with inexpensive parts.
6. The valve is relatively insensitive to air line lubricants or to other materials passing through the valving member.

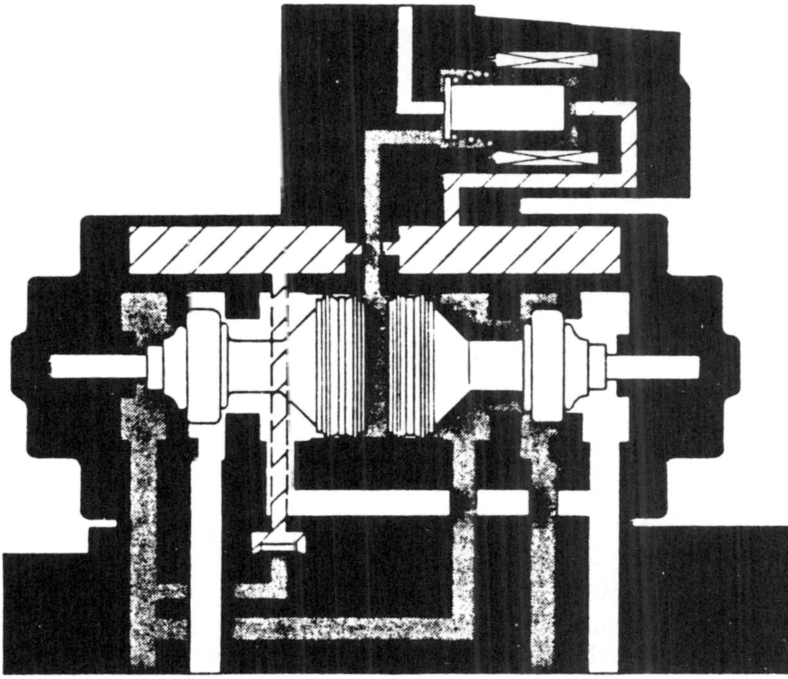

Figure 3.8

To minimize circuit problems with poppet-type valves, consider these points:

Poppet designs with directly operated actuators may be bulky in large flow port sizes.

It is difficult and expensive to obtain poppet valves in some flow path configurations.

The poppet valve design allows flow from the pressure port to escape to the atmosphere as the valve shifts. Such cross-over flow may drop the pressure at the inlet port of the valve below the rated minimum operating pressure, causing the valve to malfunction. This shift malfunction has been

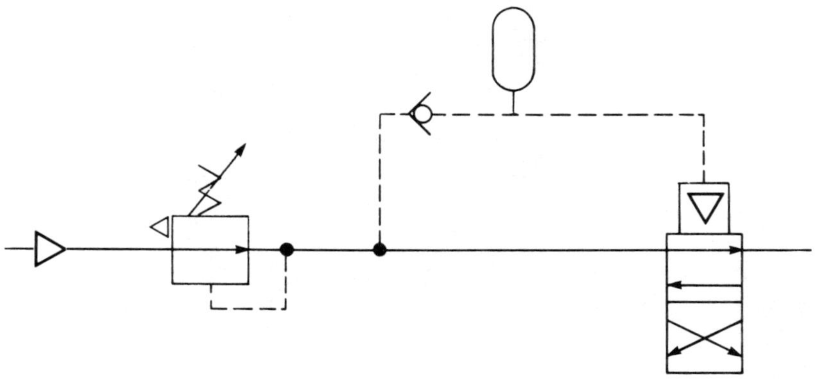

Figure 3.9

overcome in some designs by providing a reservoir of air under pressure for the pilot air circuit. This is accomplished by building in an accumulator (a miniature air receiver tank) in the pilot air supply and isolating this air reservoir from the main flow with a very lightly loaded check valve (Fig. 3.9).

3.4. DIRECTIONAL VALVE ACTUATORS

We have seen that a movable valve member can be positioned in any of its designed positions. The member is moved into these positions by manual, mechanical, electrical, or pneumatic means (Fig. 3.10). Directional valves whose spools are moved directly by muscle power are known as manually operated or manually actuated valves. Types of manual actuators include levers, push buttons, and pedals. Manual actuators are used on directional valves whose operation must be sequenced or controlled at an operator's discretion.

A very common type of mechanical actuator is a plunger. Often equipped with a roller at its top, the plunger is depressed by a cam, which is attached to the moving machine member. Mechanical actuation is used when the shifting of a directional valve must occur as an actuator reaches a specific position. Elec-

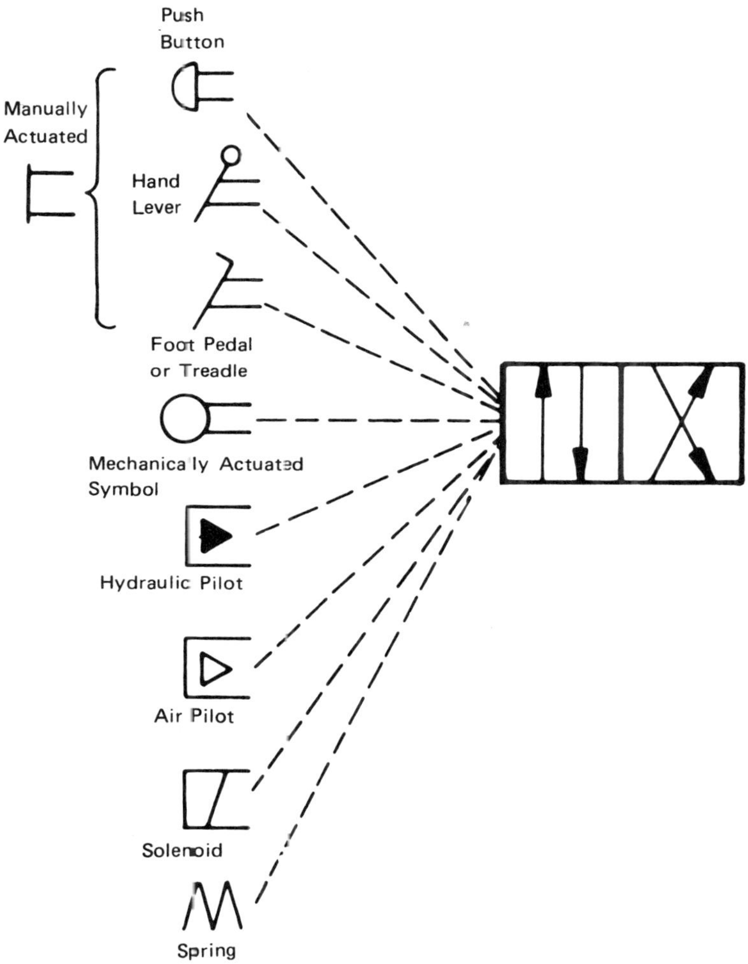

Figure 3.10

trically actuated valves are typically shifted by a solenoid in response to an electrical command.

Valves can also be shifted with air. In pneumatically actuated valves, pressure is applied to the spool ends or to separate actuator pistons.

3.4.1. Direct- Versus Pilot-Operated Valves

Where the actuating force is applied directly to the valving member(s), the valve is referred to as a direct-operated valve. When the actuating force controls a pilot fluid which in turn operates the movable valving member(s), the valve is called a pilot-operated valve. Pilot-operated valves generally require relatively small external shifting (control) forces or signals.

3.4.2. Solenoid Operation

One of the most common ways of operating a directional valve is with a solenoid (Fig. 3.11). A solenoid is an electrical device which

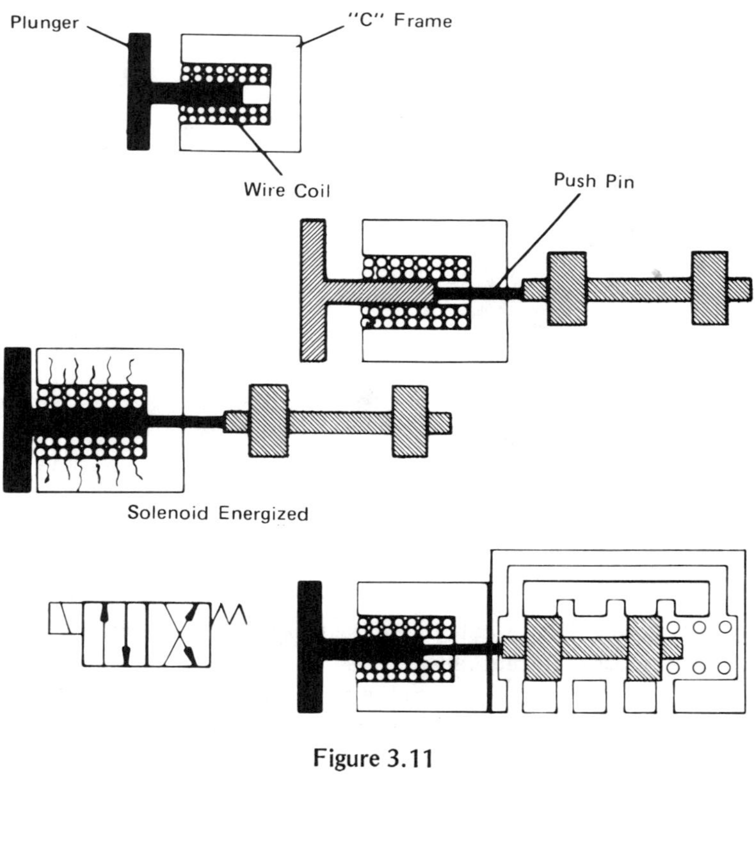

Figure 3.11

consists basically of a plunger, a copper wire coil, and a fixed magnetic assembly (frame) in which the plunger moves. The coil is usually wound on a bobbin, which is then installed in a magnetic frame. The plunger is free to move inside the coil and the fixed magnetic frame.

When electric current passes through the coil of wire a magnetic field is generated. This magnetic field attracts the plunger and pulls it into the coil, usually toward a seat, to close the magnetic path. In a valve assembly, the plunger acts against a movable member of the valve—usually the spool of a direct-operated valve or a plunger of a pilot-operated valve. In most pilot-operated valves of current design, the plunger working in the magnetic path is also the moving poppet element of the pilot valve. Signal power levels in such valves are usually under 10 watts. In a direct-operated valve, the solenoid and plunger are quite large, generally requiring power levels well above 10 watts. As the plunger moves in it can either cause a spool to move or seal off a surface, changing the flow condition.

3.4.2. Direct-Acting Solenoid Valves

When the motion of the solenoid is directly coupled to the shifting mechanism of the valve, it is called a direct-operated solenoid valve (Fig. 3.12). Two-position valves with single solenoid actuators usually depend upon a spring or air pressure to return the flow-directing element to its normal conditions. Three-position direct solenoid-actuated valves rely on large solenoid forces to shift the valve member to either extreme position from the center. Centering is usually achieved by spring forces or pressure unbalance.

Direct-operated solenoid valves offer these advantages:

1. The valve will assume the position corresponding to the actuation signal, regardless of the input pressure within the valve or circuit (provided that the maximum is not exceeded). No minimum operating pressure is required.
2. The valve remains in position even if the pressure fluctuates to low values.
3. The valve will not tend to malfunction due to restrictions or failures in supply or exhaust pressures.

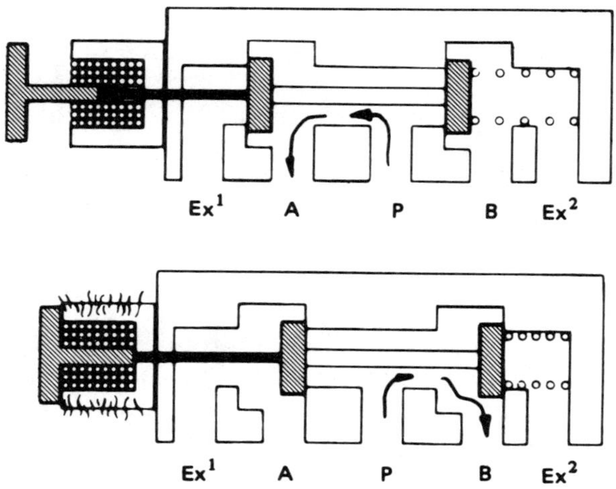

Figure 3.12

To minimize application problems with direct-acting valves, consider the following:

1. The electric power source must be capable of providing the required inrush and holding currents.
2. If actuation signals are simultaneously imposed on both solenoids of a double solenoid valve, solenoid burnout will occur.
3. Valve "sticktion" or low voltage can result in solenoid burnout.

3.4.4. Pilot Pressure-Actuated Valves

Another type of valve actuation is through a pressure pilot actuator (Fig. 3.13). A pressure pilot-actuated valve uses air pressure to move the main valve spool. This air pressure may come from a variety of sources.

Pilot-actuated valves are often used where actuation is required in remote locations. They are also useful for controlling low pressures and are a requirement in pressure-centered valves. Since actuating forces increase with increasing pressure to the

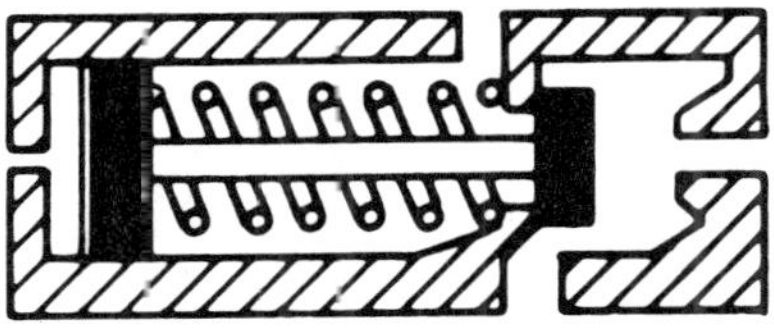

Figure 3.13

actuating ports, high forces can be generated to shift the valve member. Pilot-actuated valves may be internally or externally piloted. A valve is considered to be externally piloted if the air pressure for shifting comes from an external source. If the pilot pressure comes from a source within the valve, it is said to be internally piloted. Let us first examine the internally piloted type of valve.

3.4.5. Internally Piloted Valves

Internally piloted valves use part of the pneumatic energy delivered to the pressure port to position the movable member within the valve body (Fig. 3.14). Such a valve has a definite mini-

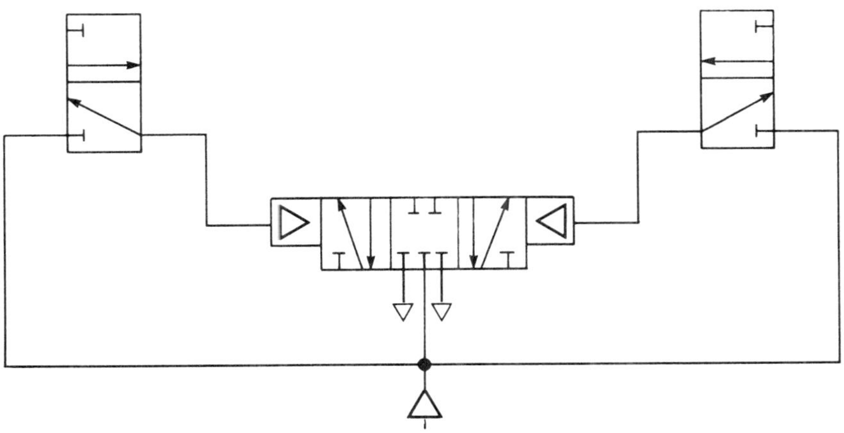

Figure 3.14

mum and typically a maximum pressure requirement. The application of this type of valve may impose circuit restrictions on the designer. This is due to the need to maintain the minimum specified pressure whenever the valve element requires shifting. If the pilot air supply exhaust line is internal to the valve, shifting may be seriously affected by restrictions in the main valve exhaust passages.

The use of properly sized, unrestricted pressure supply and exhaust lines to the valve and the pilot will normally avoid problems with valve shifting. Conversely, the use of a supply line that is too small or has built-in restrictions may cause the pressure at the valve to fall below the minimum specified operating pressure as the valve is operated. Examples of such restrictions are too small in inner diameter in fittings or hose couplings, quick disconnects, and sharp bends or kinks in conductors, expecially flexible hoses. Problems caused by such conditions may be eliminated by preventing or removing the restrictions, or by selecting a valve that is not sensitive to such conditions.

3.4.6. Externally Piloted Valves

Conversely, externally piloted valves require a pressurized source that is external to the valve (Fig. 3.15). This means that a pressure can be selected in such a way that the valve will energize and deenergize in the same amount of time. If slow shifting is required,

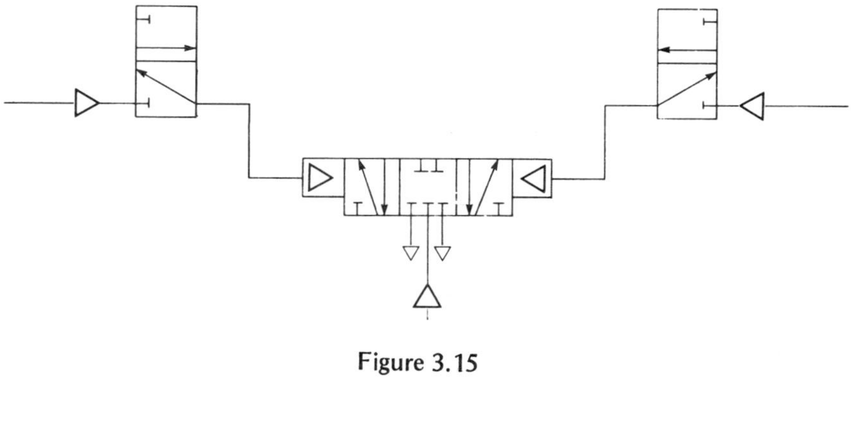

Figure 3.15

a low pressure may be used. If rapid shift rates are necessary, we can employ high pressures or the rapid exhausting of one of the pilot ports. However, when using high pressure to shift the main valve member, caution should be exercised. Extremely high pilot pressures may cause high impact forces on the valve member, leading to reduced life.

3.5. SOLENOID-CONTROLLED, PILOT-OPERATED VALVES

Forces required to shift large directional control valves directly by the actuator may be relatively high, especially if flow forces are unbalanced. To directly operate a valve of this type, large solenoids with high inrush current demands would be necessary. To oversome this problem, many designs make use of solenoid-controlled pilot operation. Such designs may provide extremely high flow capability.

Pilot-operated valves have these advantages:

1. Alternating current solenoids can be small, low-power devices, yet can operate at high inrush and low holding current levels. The result is a device with low power draw. Because the solenoid responds quickly, less heat is generated, resulting in a greatly improved relay life as well.
2. Low temperature rise at the solenoids decreases the likelihood of solenoid burnout.
3. Small solenoids can be used. This decreases the destructive "pounding" which is often the cause of mechanical failure in solenoids.
4. Solenoids operate small valve elements, which greatly decreases the possibility of failure due to varnish, dirt, and so forth.

4

Pneumatic Circuitry

4.1. INTRODUCTION

After an individual has been designing and/or maintaining pneumatic systems for any length of time, he or she will begin to notice a similarity in the circuitry. One reason for this is that many circuits share a common basic format. These basics have been time-tested and found to work quite well as a foundation for many circuits. Utilizing these basic circuits shortens the design time. This approach means that the designer does not have to start at point zero with each circuit design.

Pneumatic power circuits may be very simple, such as a single actuator controlled by a single directional control valve. On the other hand, they may be quite complex, with a multiactuator circuit utilizing directional, pressure, and "flow" control devices, which may be sequenced pneumatically or programmed mechanically, electrically, or electronically.

Today, electronic programming has become the control logic of choice for many new applications. However, many basic pneumatic circuits are on existing machines and in less complex equipment. These circuits may continue to be used for some time because they are simple, reliable, and inexpensive. It is to the circuit designer's advantage to be able to understand and use these basic systems.

The purpose of this chapter is to provide a variety of circuits that can be used as a practical foundation for a given circuit requirement, a foundation that can be built on or modified to meet all of the system requirements. In reviewing a pneumatic circuit it should be noticed that in most cases a pneumatic signal is turned on, turned off, delayed, or converted from a momentary to a maintained signal or vice versa.

The first circuit that will be examined is one used to convert a momentary signal to a maintained signal.

4.2. CONVERTING A MOMENTARY SIGNAL TO A MAINTAINED SIGNAL

Before this circuit is described the terms momentary and maintained should be defined.

Momentary is defined as the short-duration application of a command signal or shifting of an air valve by an external force. An example of a momentary shift command of a valve would be an operator depressing a push button or operating a hand lever for a short period of time (providing a pulse signal). This type of signal could be used to start or stop a sequence.

A maintained signal is defined as a command signal that, once applied, remains applied for an extended period of time, or until the shifting force is removed or overcome.

Figure 4.1 illustrates the basic momentary signal to maintained signal conversion circuit in the initial or at-rest condition. In the

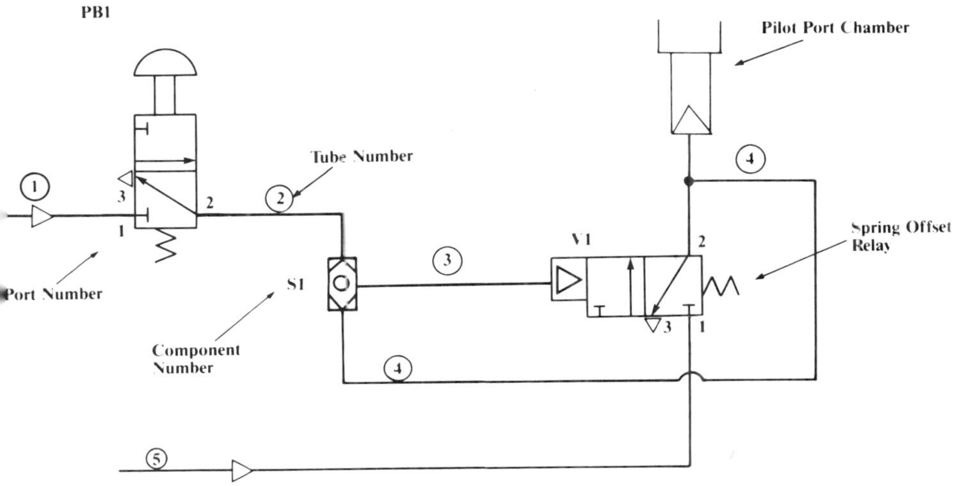

Figure 4.1

circuit, the spring offset air-piloted valve (V1) is connected to a push-button valve (PB1) through a shuttle (S1) to a chamber representing a pilot port on some device (actuator) or air valve.

Figure 4.2 illustrates the signal conversion circuit with the operator depressing the push button (PB1). Air from PB1 is directed through S1 and via line 3 to actuate V1 (valve 1). Supply pressure in line 5 is directed through V1 as the flow path from port 1 to port 2 is now passing. This pressurizes line 4 and the pilot chamber. Line 4 is also pressurized to the other port (1) of the shuttle.

Note: To aid the reader in visualizing the circuit operation, we have used hatching (diagonal lines) to indicate that a component is or has been pressurized.

The result of the operator releasing PB1 is shown in Figure 4.3. Notice that the air in line 2 exhausts through port 3 of PB1. However, the pressure present in line 4 immediately shifts the shuttle and closes off the other input and seals the pressure at pilot port A of V1.

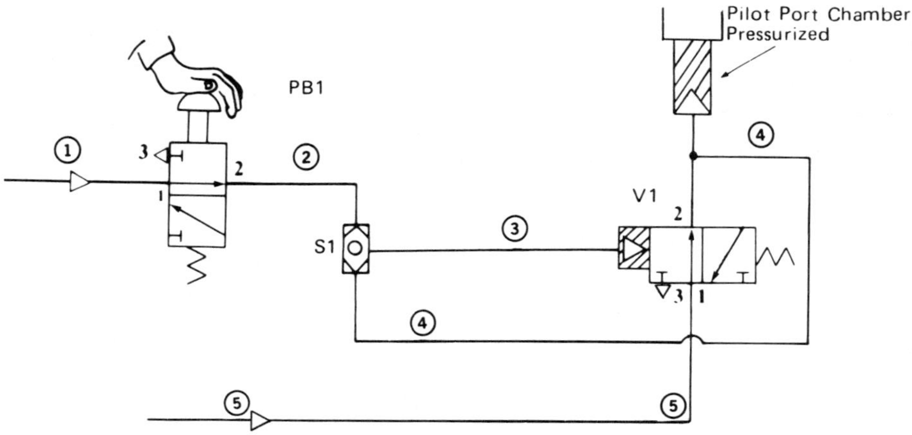

Figure 4.2

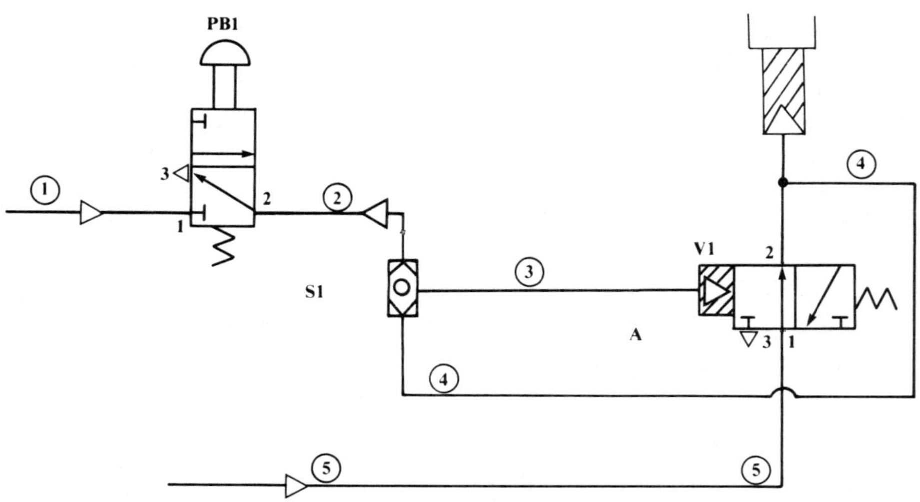

Figure 4.3

What has been accomplished is that a momentary signal (the operator depressing the push button) has been converted into a maintained signal. The signal from PB1 is said to be "memorized" by the signal at line 4—in effect, remembering within this circuit the fact that the operator has pushed the button.

If the force resulting from the pressure in line 5 should decrease below the force of the valve's spring, the pressure in line 4 would be exhausted. This function is called an "off-return" memory. If the supply pressure is lost, the relay remembers its neutral (no signal) position (because of its spring) and returns to its original condition.

This circuit network can be used to memorize a momentary signal and hold its output until it is desired to remove it. To effect removal of the output in Figure 4.3, it is necessary to vent line 3 as shown in Figure 4.4.

When the limit valve (LV1) is actuated by the cam, it vents line 6, which allows V1 to shift back to its original position, venting line 4 (Fig. 4.5). When limit valve LV1 (Fig. 4.6) is released,

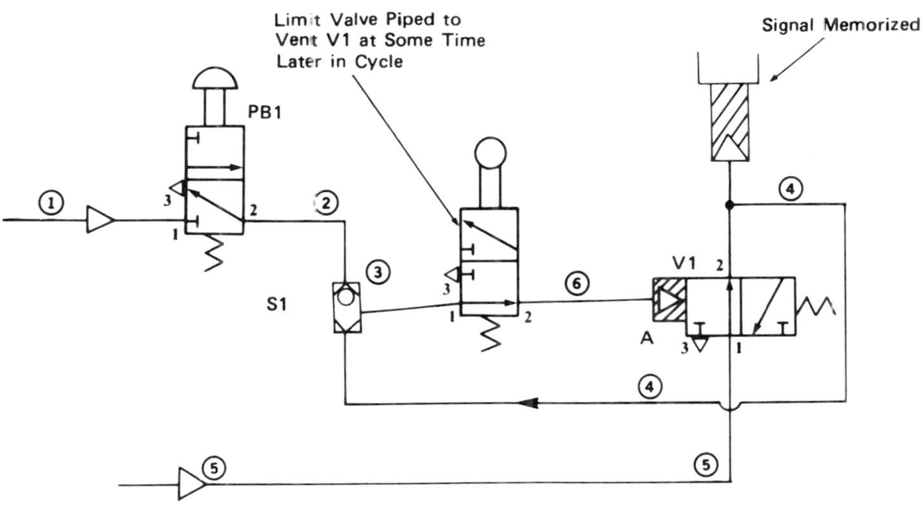

Figure 4.4

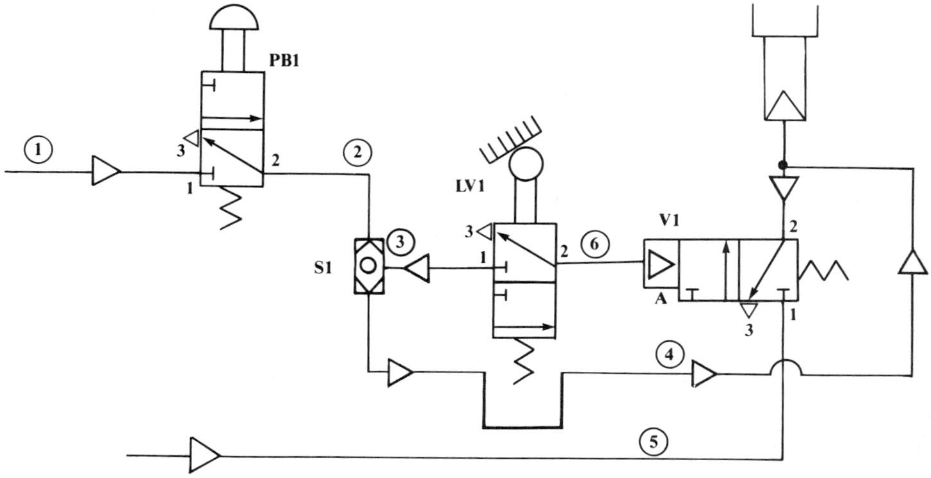

Figure 4.5 LVI made by its cam exhausts V1.

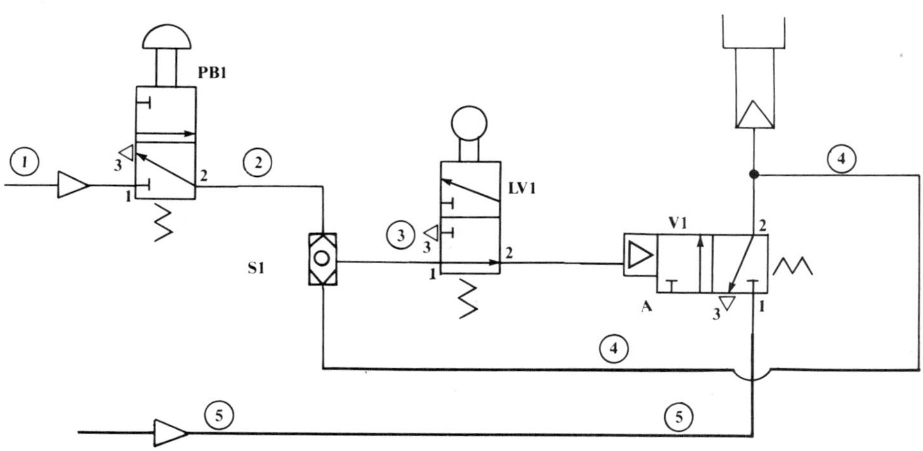

Figure 4.6 Circuit is reset, ready for next cycle.

the circuit will be back to its initial condition, ready for the next cycle. Also, a momentary signal can be converted to a maintained signal by using a spring offset valve without the shuttle. To accomplish this the three-way, two-position valve (PB1) must be replaced with a five-ported, four-way valve. A port must also be plugged. The initial condition of such a circuit is shown in Figure 4.7. PB1 is connected as a diverter function to replace the shuttle required in the previous circuit. The results of the operator momentarily depressing PB1 are shown in Figure 4.8. Notice that without a plug in port 4 of PB1, pressure in line 4 would be vented to the atmosphere. When the operator releases PB1, V1 is sealed in via the 3-to-2 flow path of PB1 (Figure 4.9). Later in the cycle, LV1 is cammed, venting line 3 out port 3 of LV1. This allows V1 to shift to its initial condition, thereby venting line 4 and line 2 out port 3 of V1 (Figure 4.10). The circuit will reset to the initial conditions when LV1 is released.

There are two disadvantages to the latter circuit. First, because port 4 of PB1 must be blocked, the normally passing flow path from port 5 to 4 cannot be used elsewhere in the circuit. Second, the use of plugged ports is not recommended because, in more complicated circuits, it makes troubleshooting exceedingly difficult and confusing.

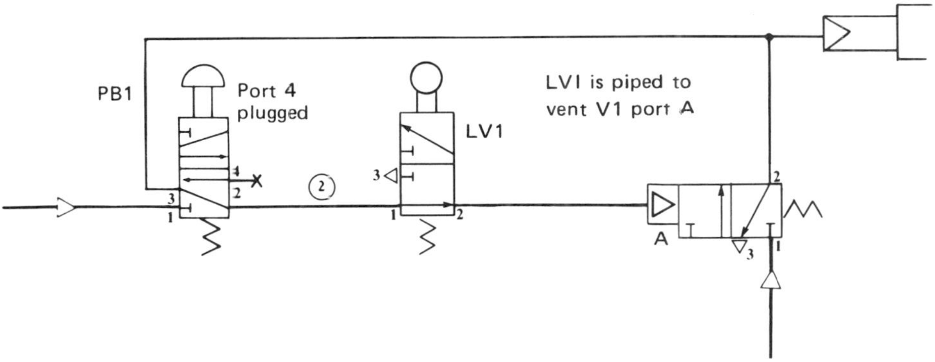

Figure 4.7

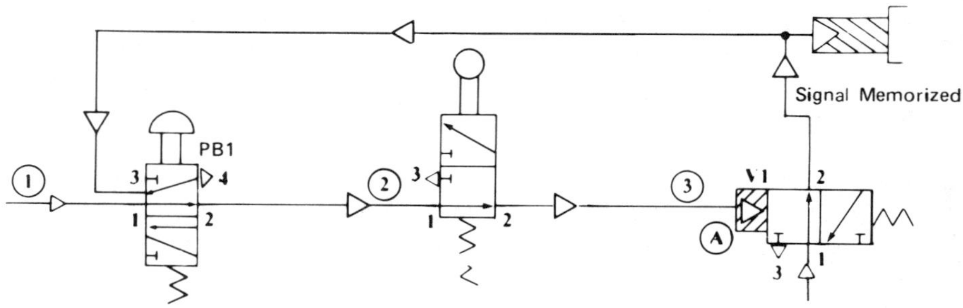

Figure 4.8

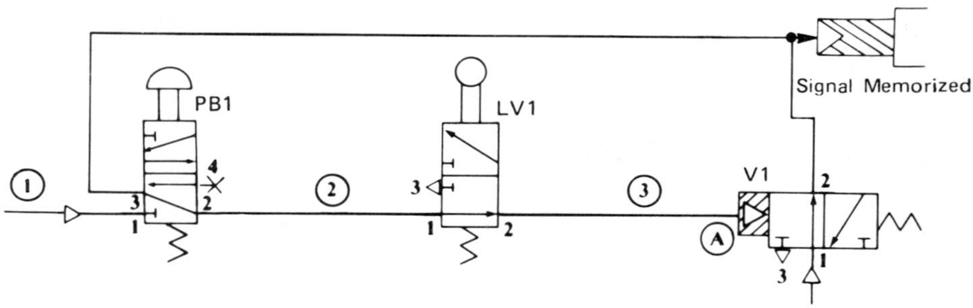

Figure 4.9

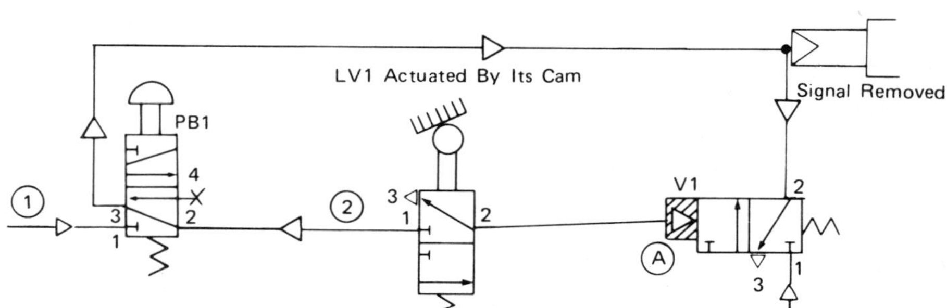

Figure 4.10

Depending on the circuit requirement it may be necessary to slow or delay a signal to a valve. Such a requirement exists when it is necessary to convert a maintained signal to a momentary signal.

4.3. CONVERTING A MAINTAINED SIGNAL TO A MOMENTARY SIGNAL

This conversion requires the use of a "flow control" valve and, if longer timing is needed, a volume changer. Separately or together they will affect the timing function.

A flow control valve (Fig. 4.11) is an adjustable orifice in a flow path with a bypass means to allow reverse flow at a low pressure drop. When air is passing through the flow control in the restricted or control direction, it must pass through the adjustable orifice, whose effective area is controlled by a metering screw. Turning the screw in one direction will open the orifice and allow more flow. Turning the screw in the opposite direction will close or reduce the orifice, restricting flow. When the air

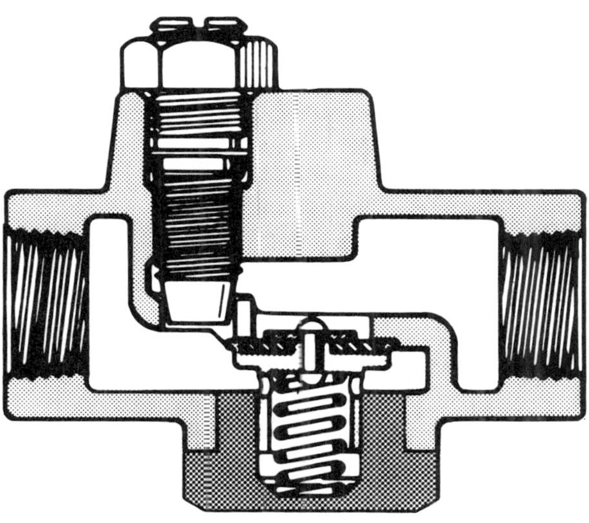

Figure 4.11 S50 shown.

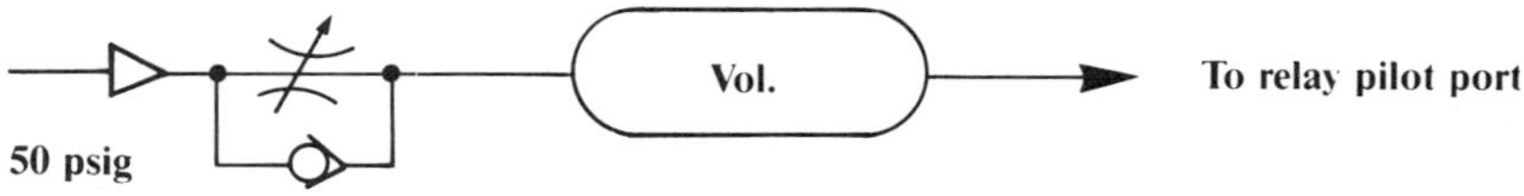

Figure 4.12 Check seated air must go through orifice.

is passing through the valve in the opposite or free flow direction it forces the bypass poppet off its seat. This allows the air to bypass the orifice. In most applications, the flow control is all that is needed to time the signal.

However, a volume chamber is connected in series with the flow control to extend the timing range of the piping arrangement. In addition, the volume chamber reduces the sensitivity of the needle adjustment by allowing a larger relative orifice setting in the flow control. Figure 4.12 shows the typical timing module. Pressure is applied to the inlet port of the flow control. The restriction in the line created by the variable orifice causes a pressure drop at the outlet port of the flow control, thus slowing the building up of pressure in the volume chamber (Fig. 4.13). Then, after a time, pressure will build in the volume chamber equal to the inlet pressure (Fig. 4.14). When the pressure at the flow control inlet is removed, the pressure dowstream of the volume chamber and the pressure in the volume chamber will exhaust around the variable orifice via the unseated check valve (Fig. 4.15).

For a given control system pressure and orifice setting it will take a given amount of time for the inlet pressure to equalize downstream of the orifice and volume chamber. If an orifice and volume chamber are located in a line feeding a pilot valve, the shifting of the valve will be delayed until the required pressure buildup takes place.

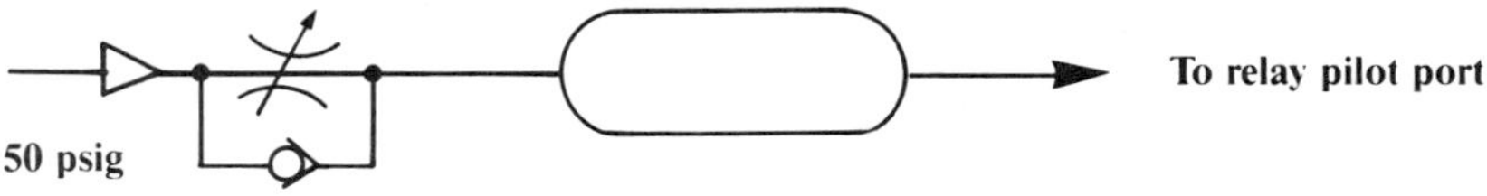

Figure 4.13 Pressure differential keeps check seated.

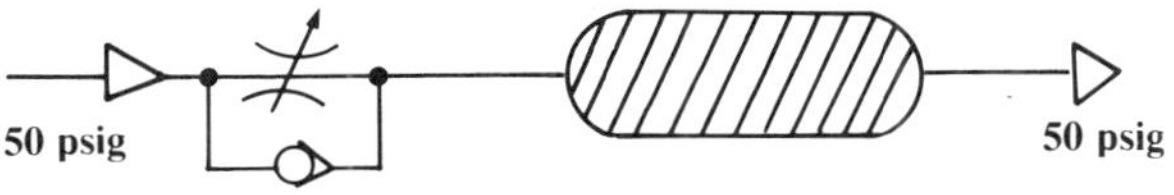

Figure 4.14 Full line pressure now on both sides of timer.

If the initial pressure and size of the orifice are kept constant, the pressure buildup will take approximately the same time for each operational cycle. If the spool's resistance to pressure is always the same, the movement of the spool and the consequent valve shift will occur with the same delay.

The size of the orifice, the pressure, or the resistance of the spool (due to a spring or air pilot) can be varied to change the range of the time delay. However, in order to keep a time delay constant, the volume of controlled pilot air must be sufficiently large that the pilot pressure remains almost constant as the operator shifts, and the orifice size for a given delay must remain constant. The smallest particle of dirt or a film of fluid (e.g., oil, water) can change the characteristics of the orifice. This is a good reason why clean, dry, nonlubricated air is needed for consistent and reliable action.

With these timing facts in mind, let us return to the "piping" arrangement required to convert a maintained signal to a momentary signal. In Figure 4.16 the spring offset valve (V1) is connected with a flow control and volume chamber (TD1) to limit valve (LV1). For the purpose of this illustration, the output of the valve (V1) is shown connected to a spring return air cylinder. Assume that the air cylinder carries a stamping pad, and the limit valve is placed at the end of a conveyor. When a box reaches the end of

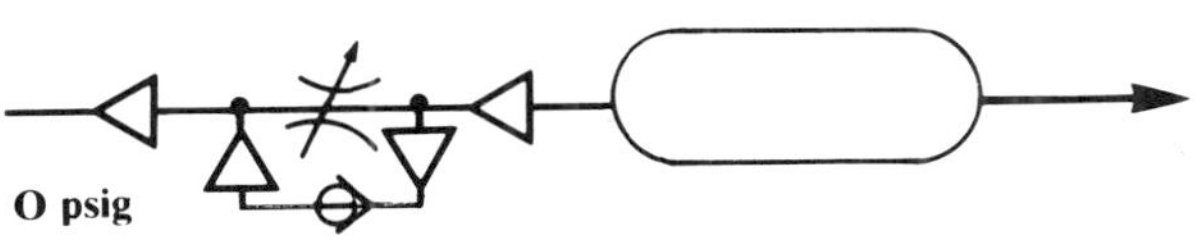

Figure 4.15 Without the check value in the flow control, the variable orifice would also delay the exhaust signal.

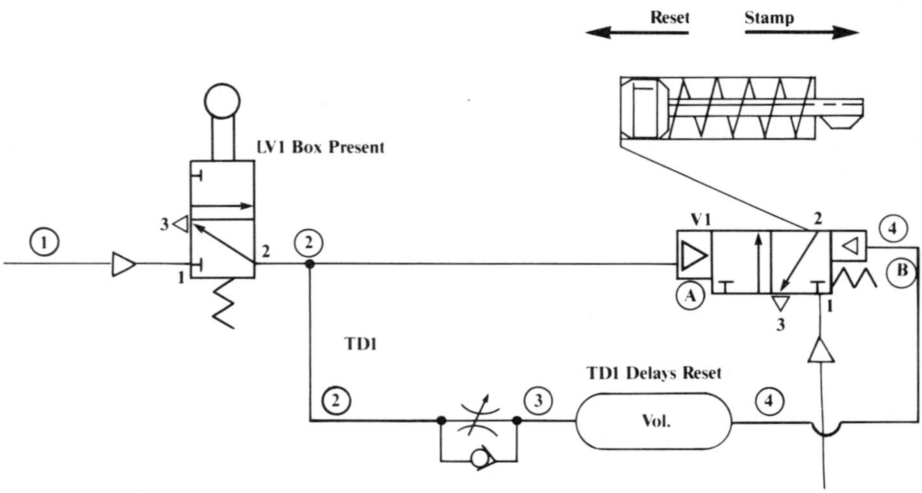

Figure 4.16

the conveyor, it is to be stamped once and is to remain in place until an operator removes the box some time later. When the box is removed, the circuit must reset for the next box. In this conversion, we will use a spring offset valve with two pilot ports. Figure 4.16 shows how the use of the B pilot port accomplishes the conversion of a maintained signal (the limit valve being held by the box) into a momentary signal (the stamp pad stamping the box only once). This function is also known as the "one-shot" function because it signals once and only once upon receipt of a maintained signal.

Figure 4.16 shows the signal conversion in its initial condition. The next figure (Fig. 4.17) shows the box making LV1, shifting V1, compressing its spring and causing the cylinder to stamp the box. At the same time TD1 (time delay) is being pressurized. After a time (when TD1 times out) the pressure in the B port of V1 builds to the spring value below the pressure in the A port (Fig. 4.18).

Now, when the pressure in the B port equals the pressure in the A port the two forces balance, canceling each other out. With nothing left to compress it, the spring forces the spool back to its

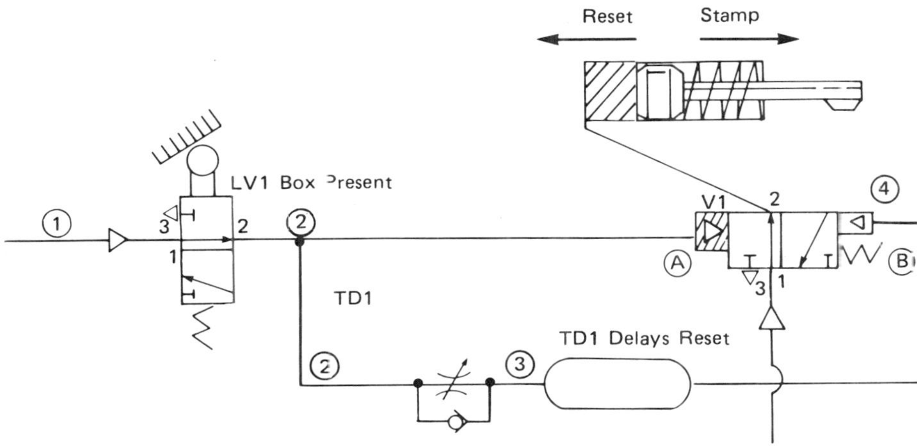

Figure 4.17

original position. This exhausts the cylinder line and resets the stamp cylinder (Fig. 4.19). When LV1 is released, lines 2, 3, and 4 are exhausted simultaneously. No movement of V1 will occur because the spring will still be in command of the valve and hold the valve in the initial position (Fig. 4.20).

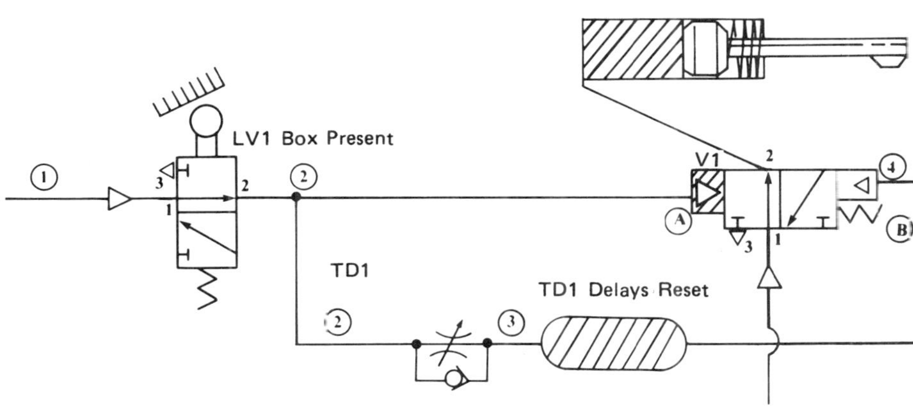

Figure 4.18

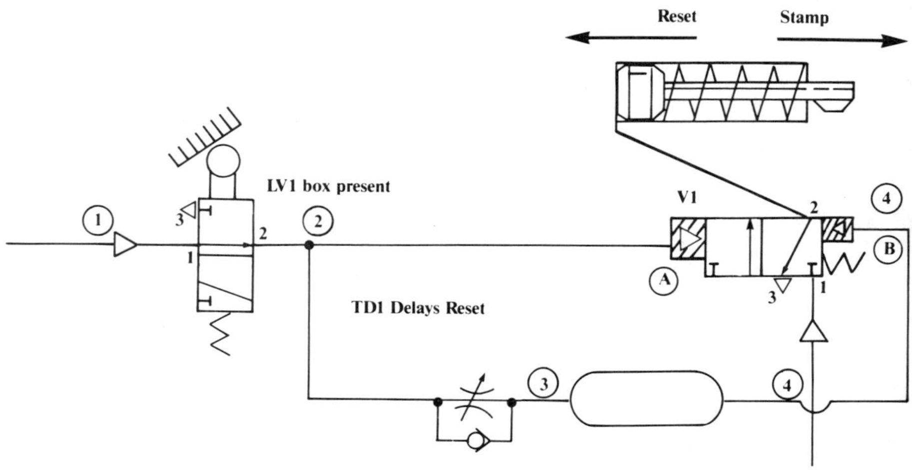

Figure 4.19

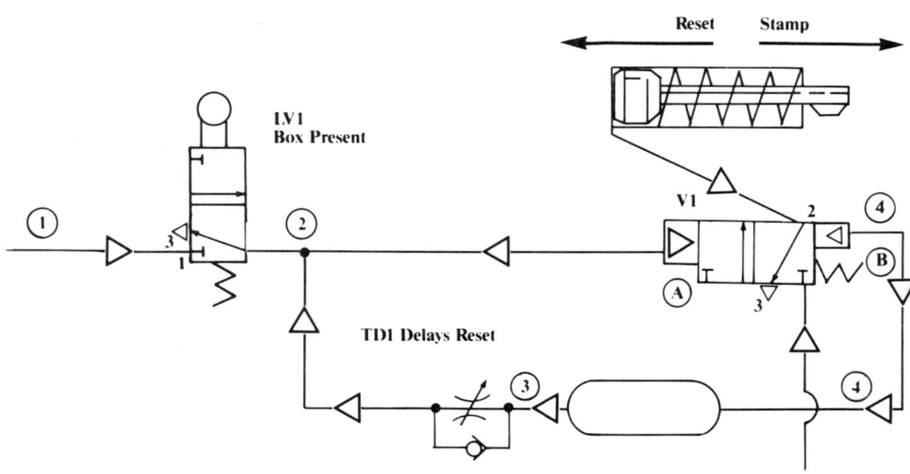

Figure 4.20

The length of time the momentary signal remains is controlled by the setting of the flow control in TD1. In the foregoing example, the cylinder immediately extended and stayed extended only until the flow control allowed the pressure to equalize on V1's spool. If LV1 was never released the cylinder would not go out again, hence the name one-shot. It works once until the maintained signal, in this case the "box present" signal from LV1, is removed.

4.4. DELAYS

Often it becomes necessary to delay a signal from an information source (limit valve) before allowing an action to occur. For example, a limit valve may indicate when a mold is closed. It is desirable to keep the mold closed for a specified period of time to allow the molded parts to cure. When the material in the mold is sufficiently cured, the mold can be opened. A spring offset valve, volume chamber, and flow control are required to accomplish this task. The circuit initial conditions are shown in Figure 4.21.

When the mold fully closes it makes LV1 and starts timer TD1 (Fig. 4.22). Next, the timer TD1 times out and the "mold open" signal is given (Fig. 4.23).

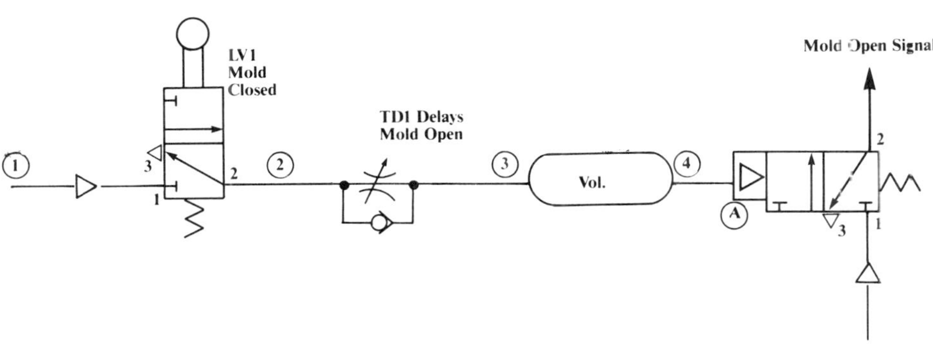

Figure 4.21

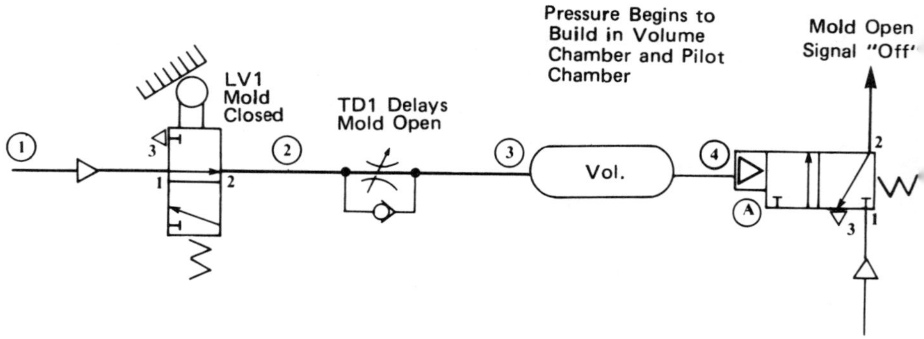

Figure 4.22

When the limit valve is released, the pilot port of V1 and the volume chamber will be exhausted simultaneously via port 3 of LV1. This will reset the circuit to the initial conditions. It should be noted that the spool is subjected to slowly increasing pressure and therefore slowly moves or creeps to its fully actuated position. This means that any valve connected to the output of V1 will also move slowly.

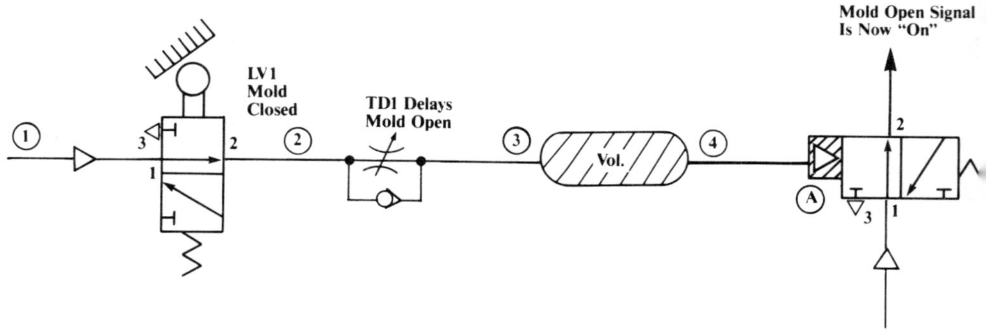

Figure 4.23

4.5. DELAYS WITH DETENTED VALVES

The detented valve is used as a memory function. A miniature five-ported, four-way valve is used for illustration. This style of valve is representative of all air valves, large or small, utilizing a mechanical type of detent assembly, often in the form of spring-loaded detent balls. Remember that many air valves rely on often highly variable internal seal friction to hold the spool in a position. Operationally, in the circuit both appear to be functionally similar.

Figure 4.24 illustrates the detented valve and its operating principles. This figures shows the valve in the initial condition. Supply (inlet) pressure is at port 3. The flow from port 3 to port 2 is passing, as is the path from port 4 to 5. Note that port 4 is exhausting to the atmosphere via port 5. It should be mentioned at this point that port numbering or lettering may vary between component manufacturers; therefore the reader's attention should be directed to function and/or flow path type.

With a momentary signal at port A the valve spool shifts. The detent balls drop into the groove on the spool shaft, thus mechanically holding the spool in the shifted position. Now the flow path from port 3 to port 4 is passing, and port 2 is exhausting to the atmosphere via port 1 (Fig. 4.25).

When the signal at port A is removed, the valve remains shifted with flow paths maintained as indicated in Figure 4.26.

A momentary signal at the B pilot port will shift the spool back to the initial condition. Then the flow path from port 3 to

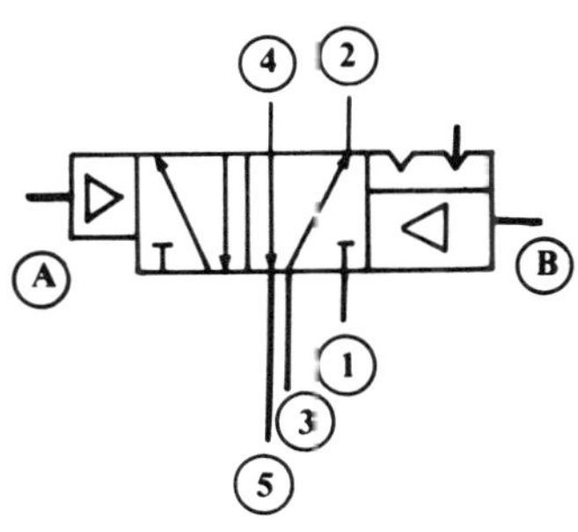

Figure 4.24

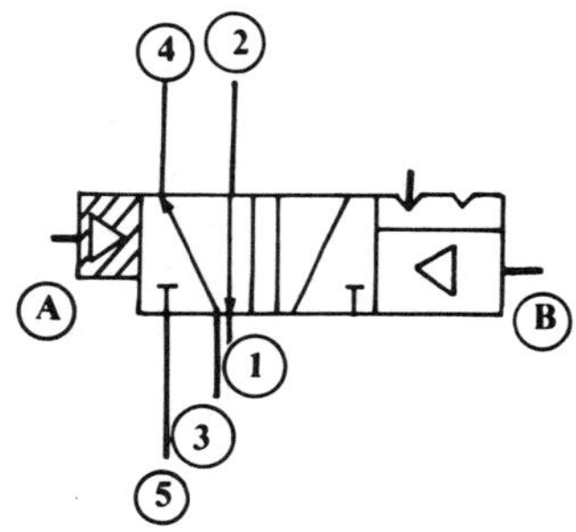

Figure 4.25

port 2 will be passing as before, and port 4 exhaust flows to the atmosphere via port 5. When the signal is removed from port B the valve spool will remain in the position shown (Fig. 4.27) until signaled again at port A.

As can be seen in Figure 4.28, the detented valve can convert a momentary signal into a maintained signal without additional components. This figure shows the initial condition.

When the operator depresses PB1 momentarily the signal is memorized in the line from port 2 of V1 (Fig. 4.29). Next the

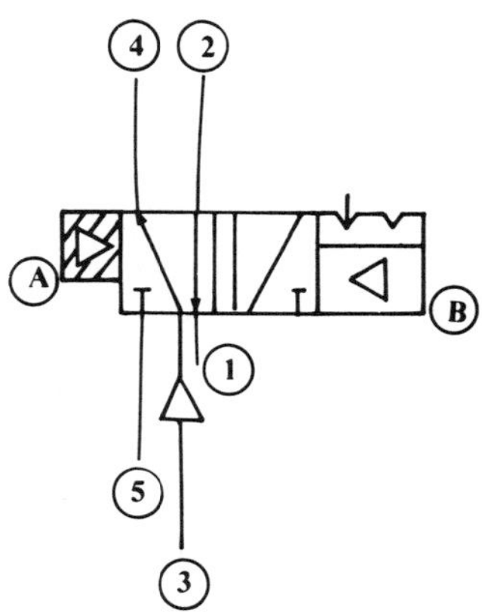

Figure 4.26

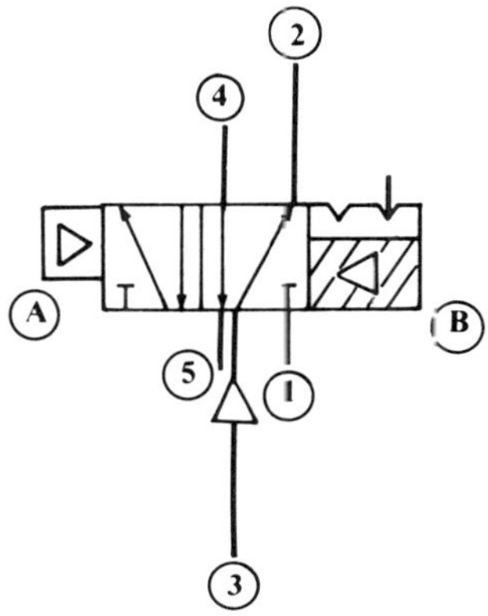

Figure 4.27

operator releases PB1, exhausting V1's A pilot port out of PB1 port 3 to the atmosphere (Fig. 4.30).

The momentary signal generated from PB1 has been converted to a maintained signal and, in effect, memorized at port 2 of V1.

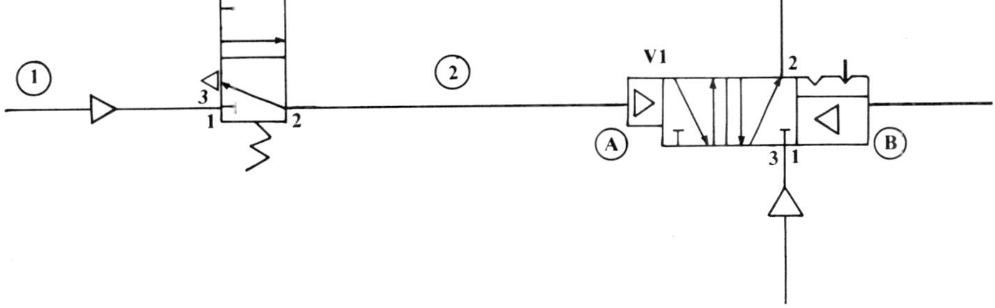

Figure 4.28

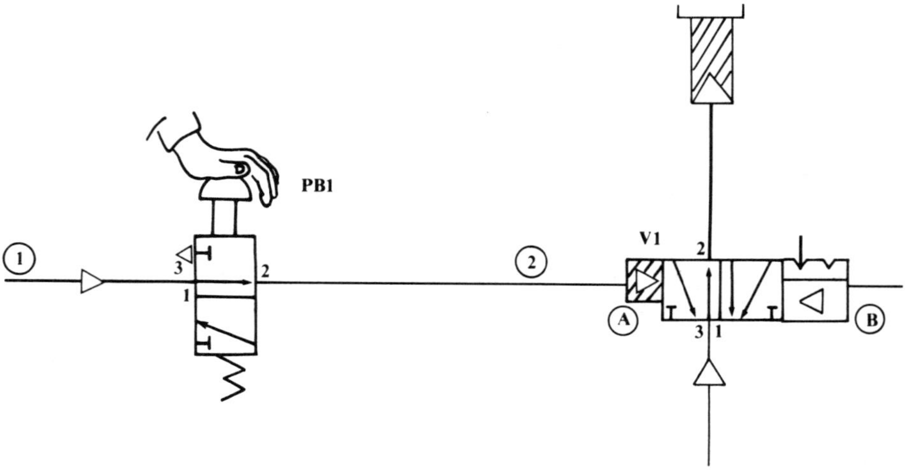

Figure 4.29

This was also accomplished in Figures 4.1 and 4.4. However, there is one basic, critical difference in this conversion created though the use of the detented relay. Should all pressure to the circuit be lost, V1 will stay exactly in the position last signaled. It remembers and holds its last commanded position, a useful feature in

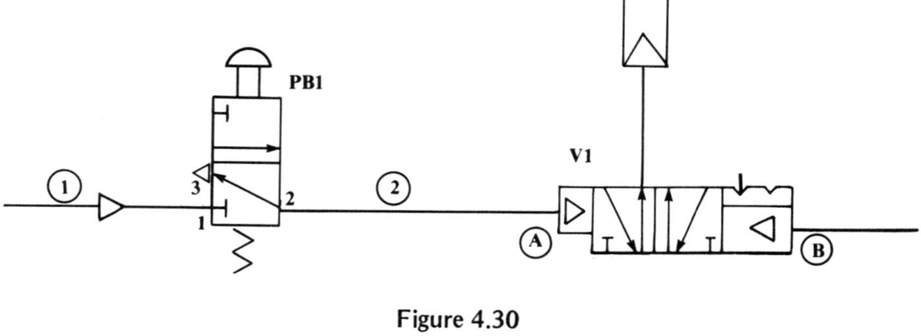

Figure 4.30

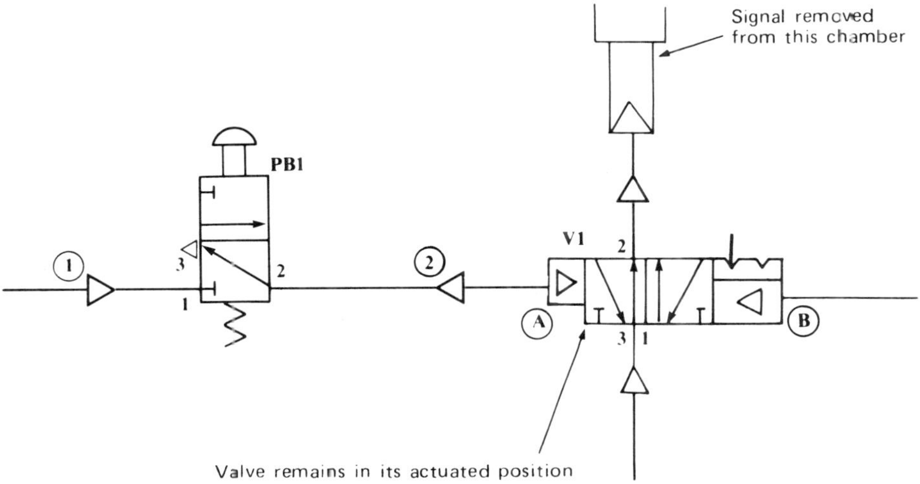

Figure 4.31

many applications. It does not possess "off-return memory" as the spring offset momentary-to-maintained conversion did. When air pressure is returned to the circuit, it will not be necessary to reactuate PB1 to energize the chamber.

Figure 4.31 shows the circuit with all air pressure removed. When pressure is reapplied, the same circuit will be pressurized, line 2 and pilot port A of V1. This function is referred to as "stabilized" memory (Fig. 4.32). In fact, the only way to remove the signal at port 2 of V1 is to countercommand the relay by pressurizing its B port momentarily, as shown in Figure 4.33. It should be noted that in order to shift V1 only one pilot at a time can be pressurized; the other must be vented. If this is not done the valve will not shift. PB2 has been added to provide a reset signal for V1.

When the operator depresses PB2 momentarily it will be reset (Fig. 4.34). When the operator releases PB2, the circuit is reset and ready for the next actuation (Fig. 4.35).

Figures 4.28 and 4.33 illustrated the action of the detented valve when signals were applied at one pilot and the opposite pilot was exhausted. The action of the spool under these conditions is a

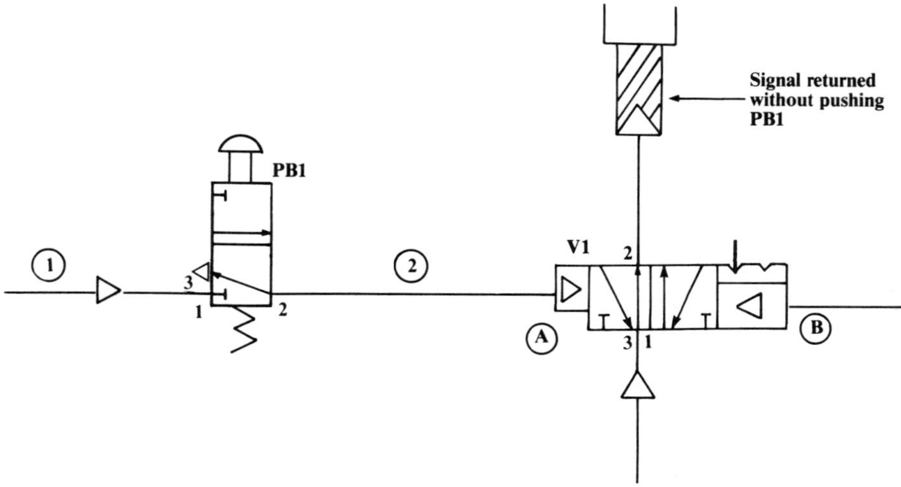

Figure 4.32

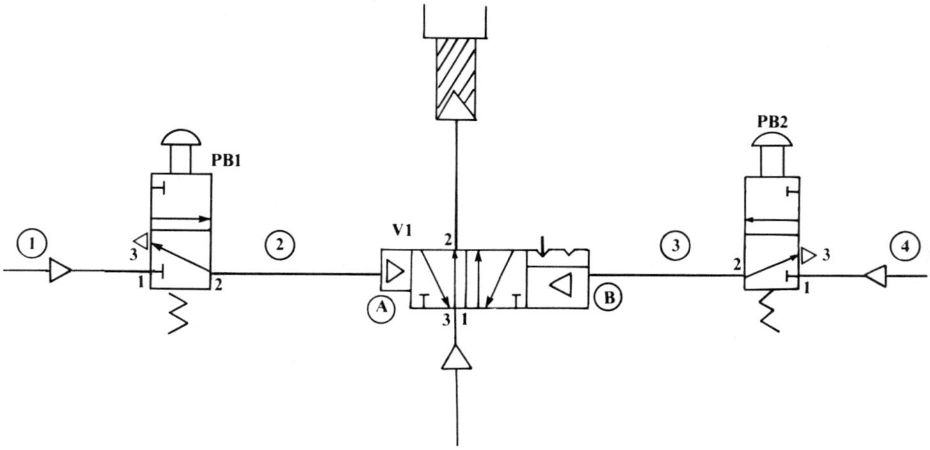

Figure 4.33

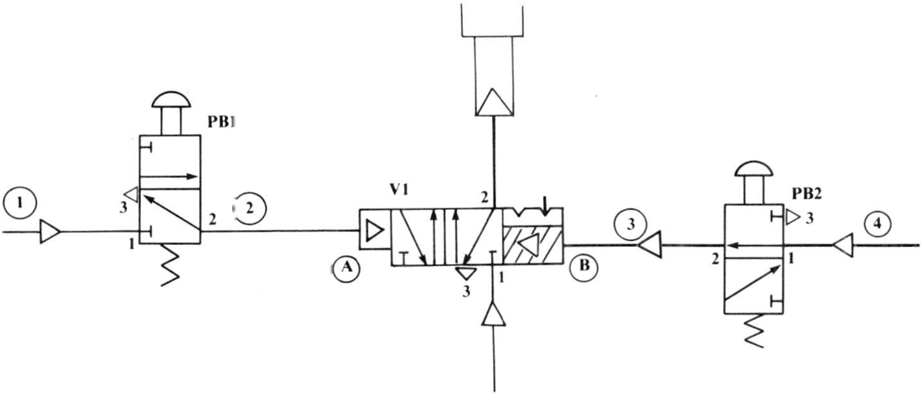

Figure 4.34

snap action because the force required to break out of the detent is far greater than the force required to shift the spool. The result is fast, predictable, and positive signal registry.

Another useful feature of the detented relay is illustrated in Figure 4.36. By actuation of PB2 the B pilot port of V1 is pres-

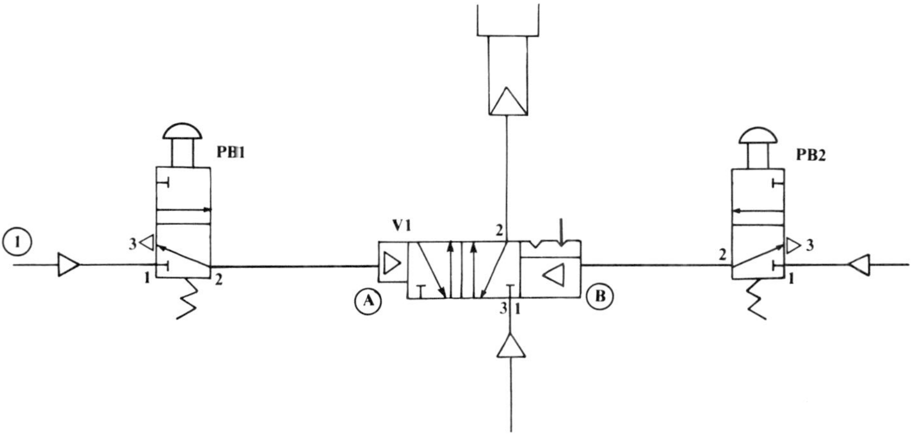

Figure 4.35

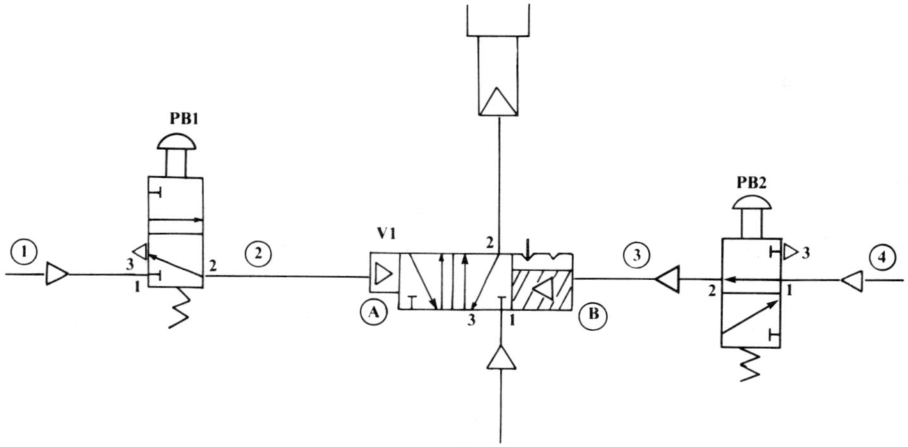

Figure 4.36

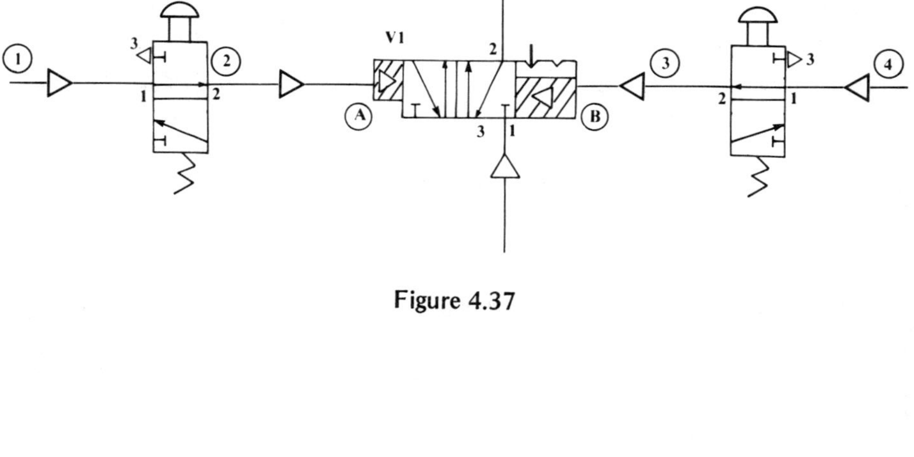

Figure 4.37

surized. Then, when the operator depresses PB1, no action or valve shift will result (Fig. 4.37).

Pressurizing both pilot ports of a pneumatic detented relay causes no damage to the valve. In fact, this technique is often used to prevent a signal from registering.

4.6. DELAYS USING THE DETENTED VALVES

Detented valves can be used to effect time delays in much the same manner described for the spring offset valve in Figure 4.21. However, the detented valve possesses two major characteristics that make it ideal for accomplishing time delays. It has the advantage of stabilized memory. This means that it produces a full on or off signal at the end of the delay. When the mold is fully closed it makes LV1 and starts timer TD1 (Fig. 4.38).

Timer TD1 times out and the mold open signal is turned from off to on (Fig. 4.39). When the limit valve LV1 is released, the flow control, volume, and pilot port V1A, will be exhausted via port 3 of LV1. The mold open signal will remain on until a reset signal is received at the B port of V1.

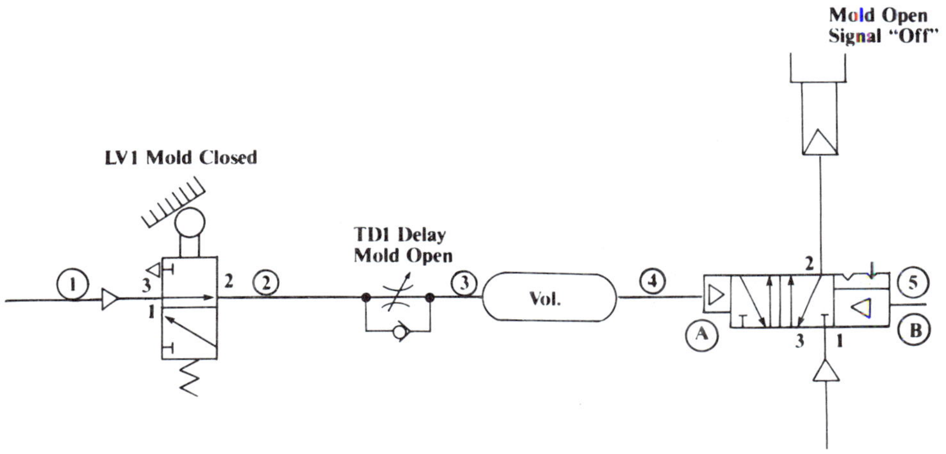

Figure 4.38

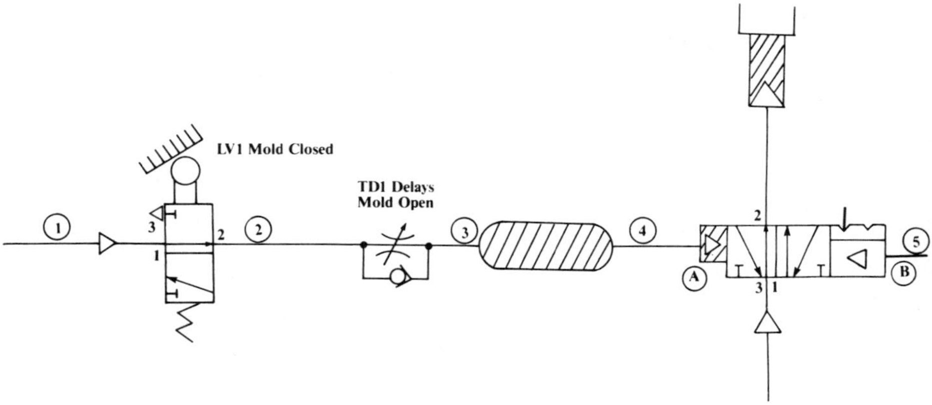

Figure 4.39

4.7. PRESSURE SENSING

Double air-piloted valves with minimal internal friction could be used in pressure-sensing circuits. This style of valve is usually piped with a well-regulated pressure at the B pilot, which acts as an air spring that causes the valve to function similarly to a spring offset valve. In such a configuration these valves are excellent pressure-sensing devices for 15–150 psig air pressure. This wide span is possible because the regulator acts as a variable spring. The regulator, by virtue of its gauge, has a direct dial readout. Figure 4.40 illustrates the initial conditions of a pressure-sensing module connected to sense the pressure at the end of a clamp cylinder. The regulator is set at 35 psig because it is desired to start the drill forward when the clamp is holding the part with at least 36 psig working on the clamp piston.

As the clamp begins to extend, the pressure in the cap end of the cylinder starts to build as shown in Figure 4.41. As the pressure begins to build in the cap end of the clamp cylinder, it is also applied to the A end of the spool of V1. Valve V1 is being held in position because of the "bucking" or "bias" pressure provided by the regulator at pilot port B.

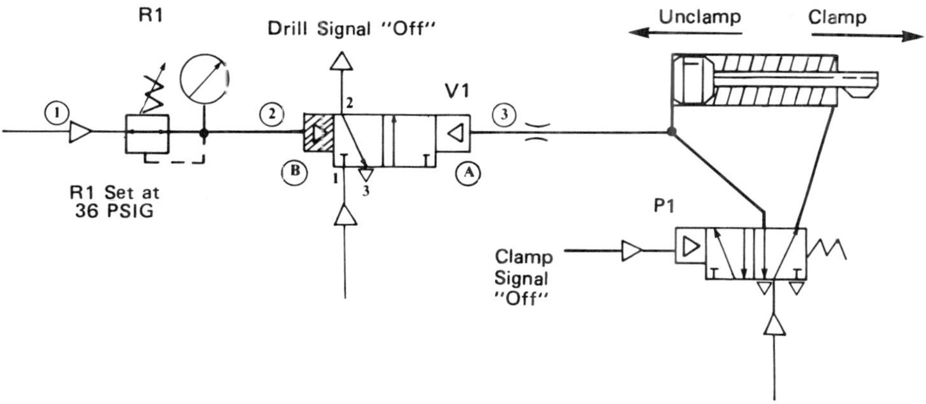

Figure 4.40

When the clamp is fully clamped, the pressure in the cap end of the cylinder will exceed the regulator setting and thus shift V1, which will turn the drill on as required (Fig. 4.42).

The set point at which the valve shifts can be easily and quickly changed by simply setting the regulator to the pressure desired.

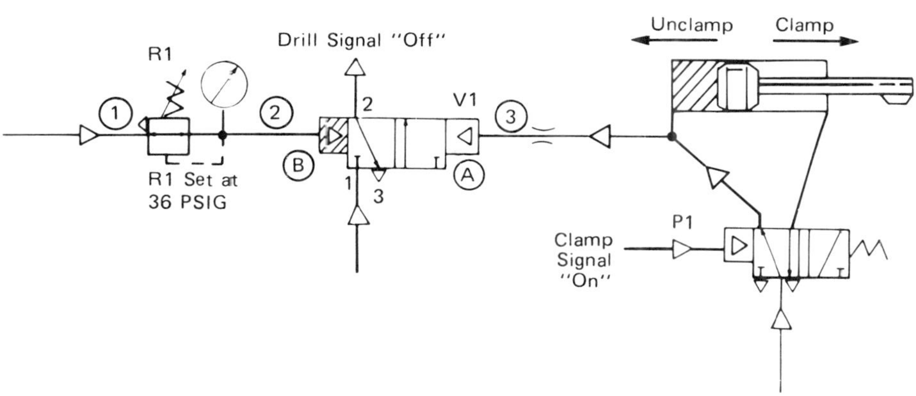

Figure 4.41

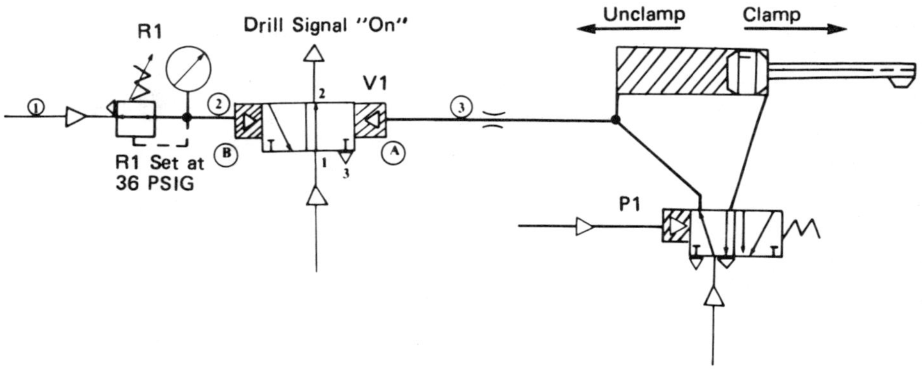

Figure 4.42

The pressure at the clamp cylinder must be higher than the regulator bias.

When the clamp signal is removed, the clamp cylinder will reset to the unclamped position. As the pressure in the cap end of the cylinder falls below 35 psig, V1 will also reset by virtue of its 36 psig equivalent "air spring." Resetting V1 removes the "drill on" signal (Fig. 4.43).

> *Caution notes:* At this point two caution notes should be emphasized. The first is that this circuit senses *only* the pressure at the cap end of the clamp cylinder. It does not know the position of the clamp cylinder. That is, if the clamp cylinder becomes blocked before it reaches the workpiece, cap end pressure will build, shifting V1 and starting the drill even though the clamp or part is not in position. Another possible problem would be the clamp cylinder becoming momentarily blocked, thus giving V1 and the drill a false signal. Perhaps the drill would then start down, then reset before reaching the workpiece. Also, if there was a restriction in the line to the cap end of the cylinder, that is, between P1 and line 3, a false signal could be generated at port A of V1, shifting the valve. This would send the drill down as the clamp was releasing the workpiece. The restriction could be a flow control, a bent or kinked line or hose, or a plugged exhaust port of P1.

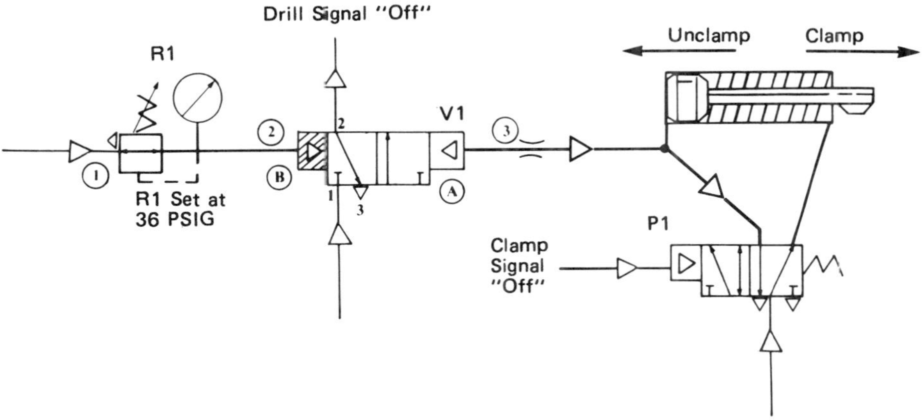

Figure 4.43

These are all unsafe conditions that must be taken into account when designing a circuit. Good design and maintenance practice would be to sense both position and pressure, that is, to add limit valves to tell whether the clamp cylinder is in position (touching the part) and holding (air pressure on the cylinder piston). Also, the line from P1 to the cylinder must be unrestricted. Careful installation and periodic inspection of this line will help ensure long and safe operation.

4.8. VACUUM SENSING

Vacuum sensing is done in essentially the manner described above except that the actions are reversed; in a sense, the spool is pulled rather than pushed. In place of regulators, variable orifices are used to let atmospheric pressure act on the spool of VV1 (vacuum valve 1) at the desired values. In Figure 4.44, a valve similar to the one used in Figure 4.40 is piped to provided information about whether a part is present on a vacuum cup.

In the example the initial conditions are as follows: no part is present on the vacuum cup; both vacuum pumps are "on" and pumping. The filters are installed to protect the valve pilot ports and the variable orifice from atmospheric contamination.

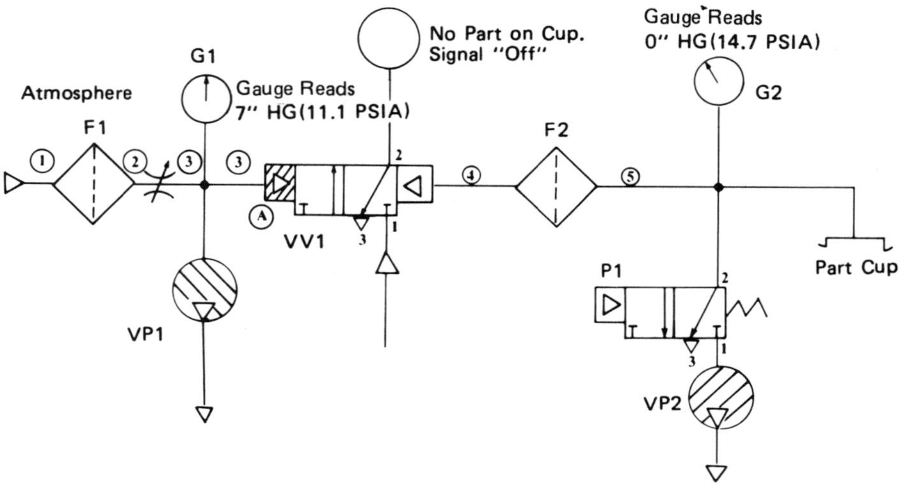

Figure 4.44 Both vacuum pumps are capable of drawing a vacuum of 20 in. HG (4.3 PSIA).

Gauge G1 reads 7 in. Hg because the orifice is allowing enough air into VV1's port A pilot to prevent vacuum pump VP1 from drawing a full 20 in. Hg. The atmospheric pressure in VV1's port B is acting as a 14.7 psi air spring. The result is to hold the spool of VV1 in a nonshifted or initial condition.

Vacuum valve P1 is turned on by an external pilot signal (Fig. 4.45). Now vacuum pump VP2 is trying to pull a vacuum while connected directly to the atmophere by valve P1 (Fig. 4.45). Since the relative size of the vacuum cup is much larger than the variable orifice, VP2 cannot pull a higher vacuum than VP1; therefore VV1 stays in the position shown. Relatively, there is higher-pressure air on the (B) end of the spool of VV1 than on the (A) end. This keeps the B pilot in command of the valve.

However, when a part is placed on the vacuum cup it allows VP2 to begin drawing a vacuum in VV1's B port, line 4, F2, and line 5 (Fig. 4.46). As pump VP2 continues to pump, VV1's B pilot is evacuated to 9 in. Hg, creating a differential on the spool. There is now a greater air spring in the A pilot and the spool will therefore shift. Vacuum pump VP2 will continue to evacuate the

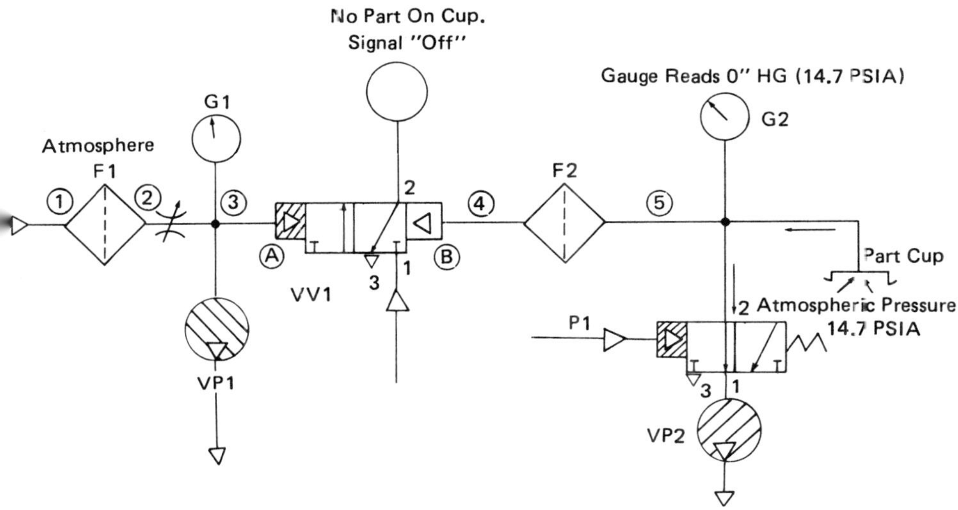

Figure 4.45

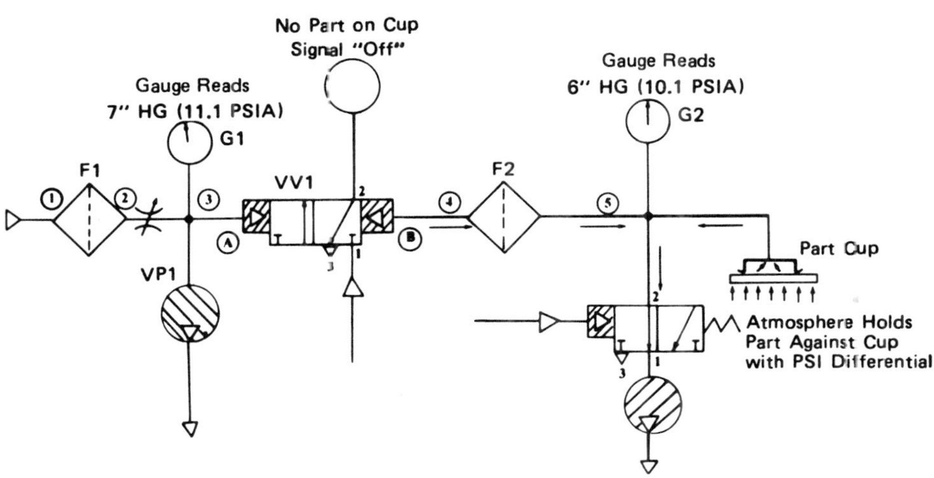

Figure 4.46

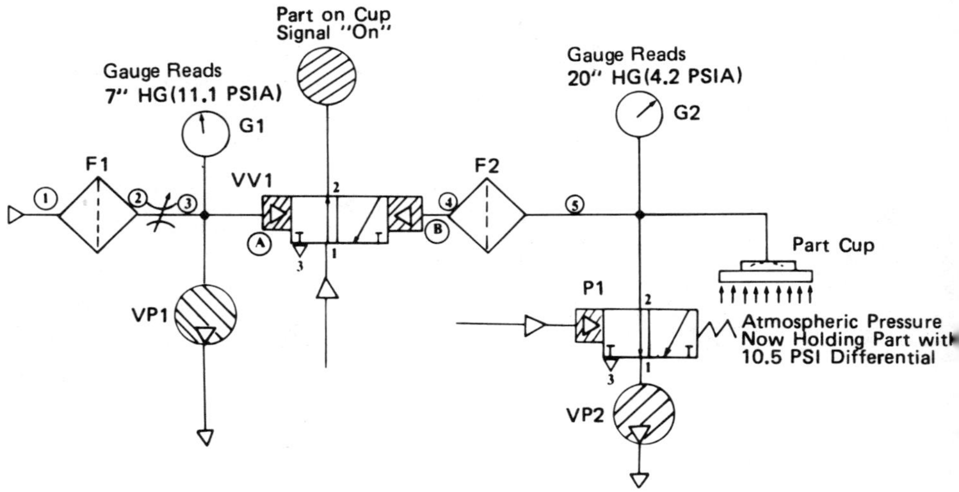

Figure 4.47

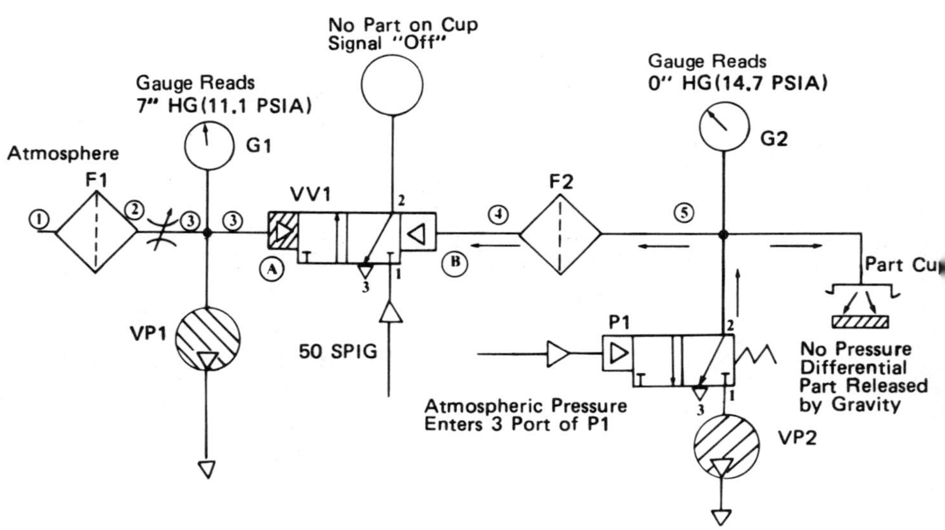

Figure 4.48

B pilot. This will create a greater pressure differential to hold the spool in the actuated or shifted position (Fig. 4.47).

Next, when valve P1 is de-energized, the part will be released and atmospheric pressure will reset VV1, removing the "part on cup" signal. Also, VV1's B pilot is again in command of the valve (Fig. 4.48).

> *Note:* Not all pneumatic directional control valves are capable of operation in a vacuum system; therefore valve manufacturers should be consulted concerning their particular valves and the valves' ability to operate in a vacuum system.

5

Valve Applications

Common Pneumatic Circuits

5.1. INTRODUCTION

There are probably not two pneumatic system designers who will generate identical circuits to perform a complex control function. However, most will agree that there are several basic or "old stand-by" pneumatic circuits. These circuits are the building blocks for many of the more complex systems in use today.

The circuit designer has the primary responsibility of carefully examining the circuit specifications, selecting the correct components, and choosing the best basic circuit to meet the system requirements, or to use as the system's foundation.

This chapter presents several of these "common pneumatic circuits," beginning with the simpler and proceeding to the more complex. As each circuit is presented its operation will be explained.

The circuits presented are intended only as a guide to meeting a system design requirement.

5.2. TWO-WAY VALVE CIRCUITRY

The two-way valve is used primarily as a shutoff valve, like the valve in a kitchen sink. The two-way does not provide an exhaust function. This valve could be used to control parts "blowoff" (i.e., the use of a puff of air to eject a part from a machine). An example would be transferring toothbrushes from final assembly to the packaging machine by blowing them through a tube (Fig. 5.1).

The valve can be used to isolate segments of a pneumatic circuit. In Figure 5.2 the air or surge tank is isolated.

The two-way valve can also be used to affect cylinder rod speed. Tubing the exhaust to a normally open two-way valve allows the cylinder to travel at high speed. When the solenoid actuates the valve, the exhaust must travel through the restriction in the flow control, reducing the cylinder rod speed (Fig. 5.3).

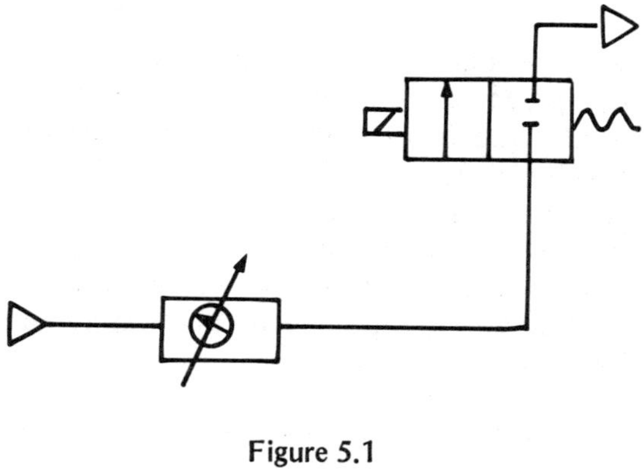

Figure 5.1

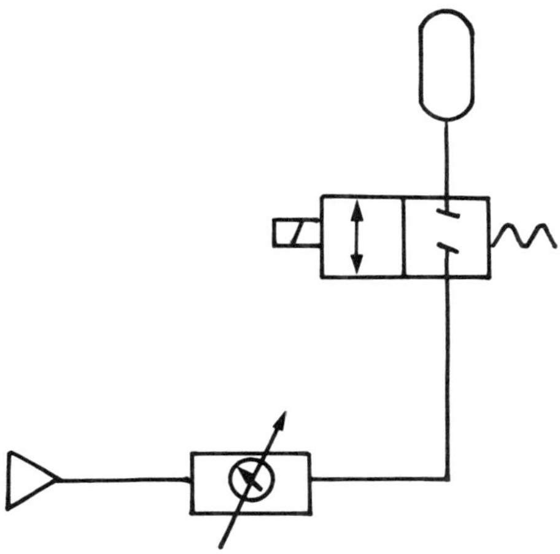

Figure 5.2

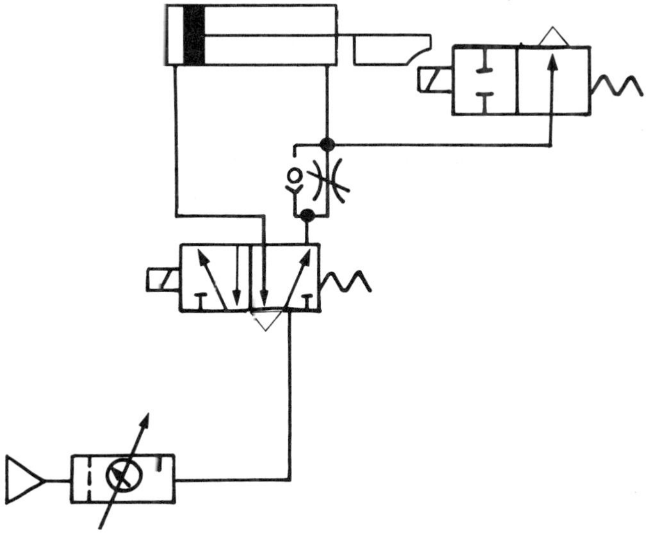

Figure 5.3

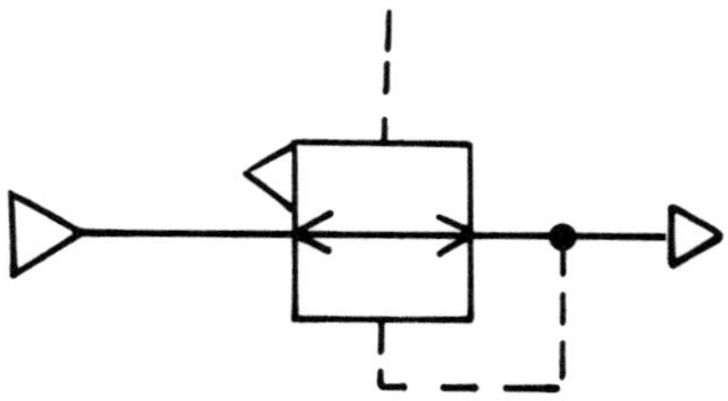

Figure 5.4

As mentioned earlier, air pilot-operated regulators can be used as a two-way valve (Fig. 5.4) and generally cost less than a valve with equivalent flow capacity. An air pilot-operated regulator can be very effective as the main on-off valve for a machine. A 1½-in. ported regulator in the air supply header can be controlled by a small solenoid or manually operated three-way valve mounted in a location convenient for the operator.

5.3. THREE-WAY VALVE CIRCUITRY

The three-way valve offers the exhaust function. It provides a means of pressurizing and exhausting a vessel (Fig. 5.5). Unlike a two-way valve, which allows only one flow path, the three-way valve provides two flow paths to the output of the valve.

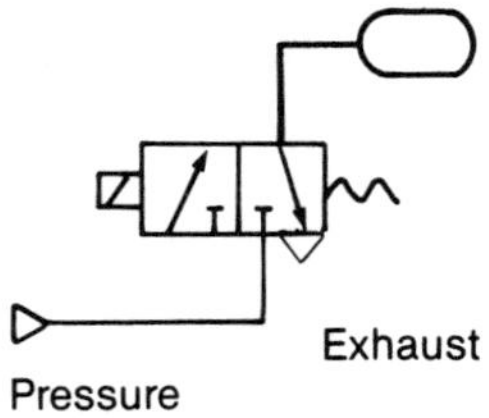

Figure 5.5

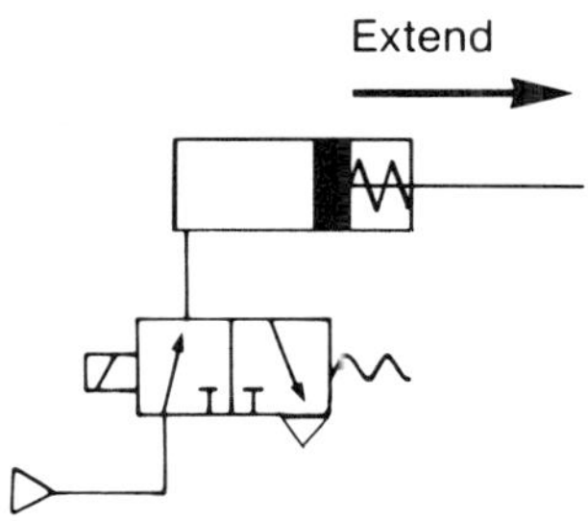

Figure 5.6

Consider a single-acting cylinder (Fig. 5.6). When the solenoid is energized, the valve shifts, pressure enters the cap end of the cylinder, the force developed overcomes the spring, and the cylinder extends.

When the three-way (Fig. 5.7) two-position valve solenoid is de-energized, the spring shifts the valve, the pressure exhausts, and the spring in the cylinder returns the piston to the start position.

Three-way valves can be used for the basic "and" circuit (i.e., two actions are required for an output). Two normally closed,

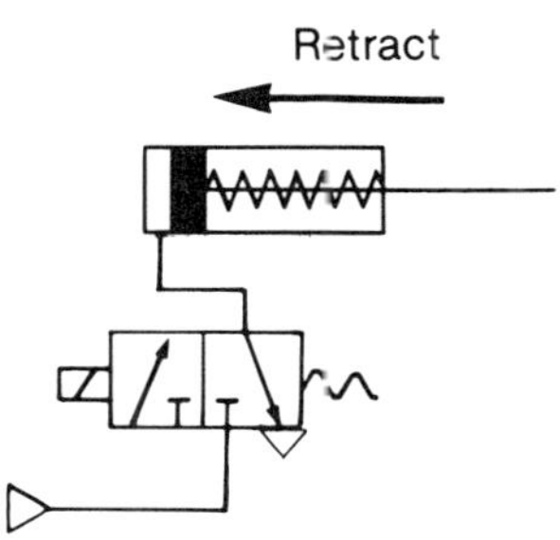

Figure 5.7

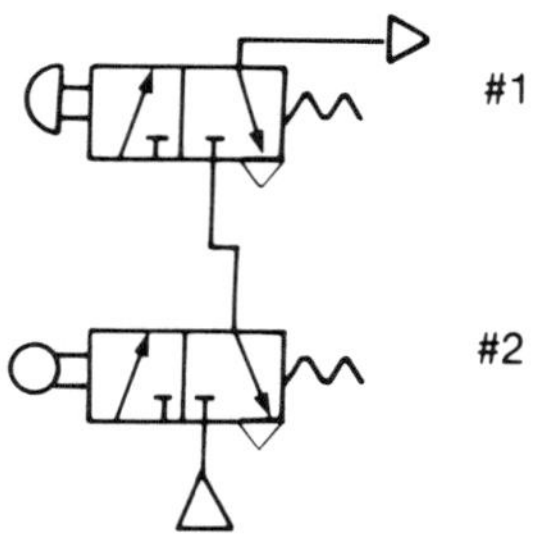

Figure 5.8

three-way valves are required (Fig. 5.8). Valve 1 could be a palm button; valve 2 could be a limit which indicates that a safety gauge is in place. Both valves must be actuated before the machine can start. Note that the circuit in Figure 5.8 does not meet Occupational Safety and Health Administration (OSHA) requirements for a two-hand, non-tie-down safety circuit.

The basic "or" circuit can be accomplished by using 2 three-way valves and a shuttle valve (Fig. 5.9). The shuttle valve allows either action to provide an output. Valve 1 could be a limit valve actuated by a carton on a conveyor. Through the shuttle valve either valve 1 or valve 2 could provide an output to start the machine.

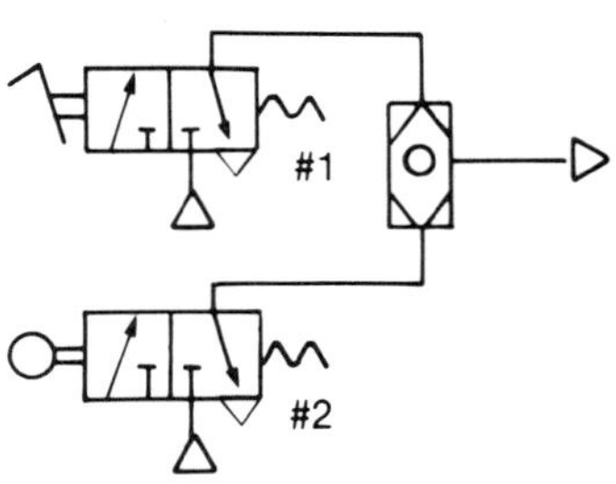

Figure 5.9

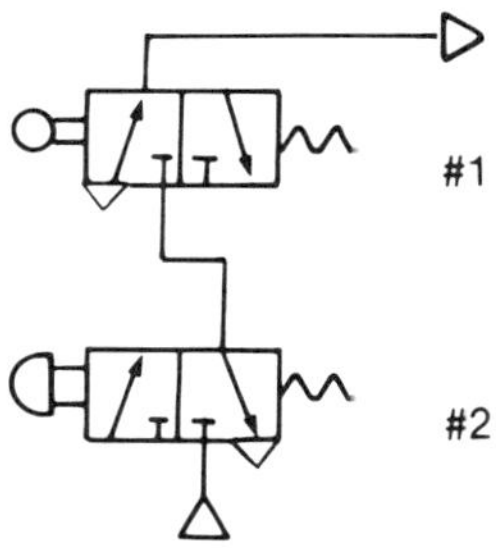

Figure 5.10

In the basic "not" circuit (Fig. 5.10) a normally open three-way valve interrupts the output when actuated. Valve 1 represents a means of monitoring a system. An example would be a full conveyor. The system will stop until valve 1 is deactuated.

The following "seal-in," "off-reset," and "retentive" memory circuits have been presented in previous chapters. However, they are shown again here as a review and to retain the continuity of this chapter on building block circuitry.

Figure 5.11 represents the basic seal-in type of memory circuit. When valve 1 is actuated, the output passes through the shut-

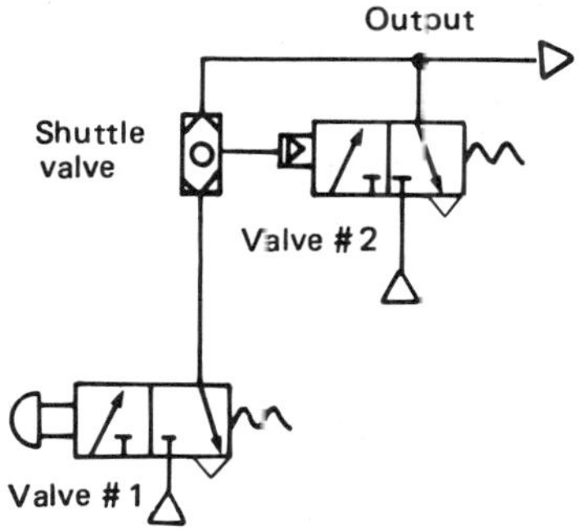

Figure 5.11

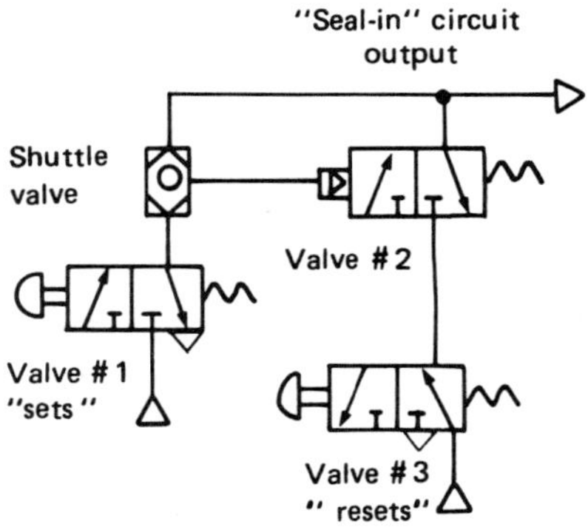

Figure 5.12

tle valve actuating valve 2. The output passes through the shuttle maintaining pressure at its own pilot port. A momentary output of valve 1 is converted to a maintained output of valve 2 (momentary-to-maintained signal conversion). This type of circuit is also known as an off-reset memory circuit. The output of valve 1 is memorized by valve 2. The memory is reset when pressure is removed from the circuit and the spring in valve 2 returns the valve to its normal position.

The off-reset memory function can be controlled by the introduction of a reset valve (Fig. 5.12). The circuit shown in Figure 5.12 uses a momentary output from valve 1 to actuate valve 2 as we described in Figure 5.11. Valve 3 is a normally open valve which when actuated will exhaust the supply from valve 2, allowing the spring to return the valve to the start position. The circuit function can be described as valve 1 sets the memory while valve 3 resets the memory.

Figure 5.13 represents a retentive type of memory circuit. A momentary output from valve 1 shifts valve 2. The double pilot valve retains its position until a momentary signal from valve 3 returns it to the initial condition. Valve 2 provides a memory which is not affected by supply pressure. The operator can shut the machine down, turn the air supply off, and when the machine starts valve 2 remains in the same position it was in when the

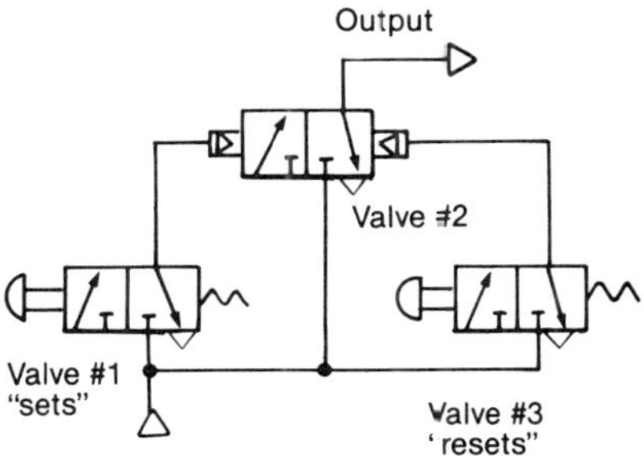

Figure 5.13

machine stopped—thus a retentive memory (momentary-to-maintained signal conversion). *Note:* for best results use a valve provided with a detent mechanism in both positions.

A type of interlock circuit is shown in Figure 5.14. Its purpose is to prevent two things from happening at the same time, to

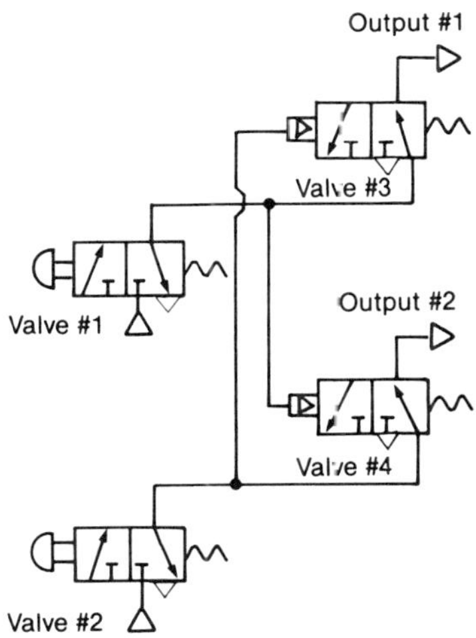

Figure 5.14

prevent damage to a horizontal and a vertical cutting tool if they were to be actuated simultaneously. When valve 1 is actuated air passes through the normally open valve 3, which provides output 1. This also actuates valve 4, eliminating any possibility of output 2. Valve 2 will operate in the same manner to provide output 2 and prevent output 1. However, when valve 1 and valve 2 are actuated simultaneously, output 1 and output 2 are both exhausted.

5.4. FOUR-WAY VALVE CIRCUITRY

The four-way, four-ported valve shown in Figure 5.15 has been actuated. Pressure is being applied to the cap end of the cylinder while the air in the rod end is being vented to exhaust. The cylinder is extending. When the solenoid is de-energized the spring returns the valve to its initial condition. Now pressure is being applied to the rod end of the cylinder while air in the cap end is being vented to exhaust. The cylinder is retracting (Fig. 5.16). A four-way, four-ported valve can move the cylinder in either direction.

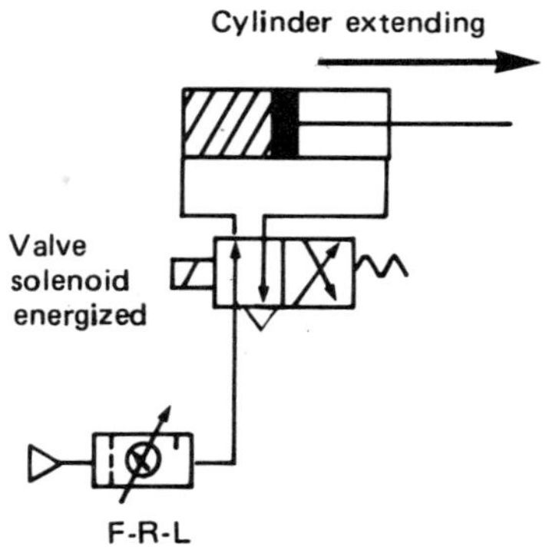

Figure 5.15

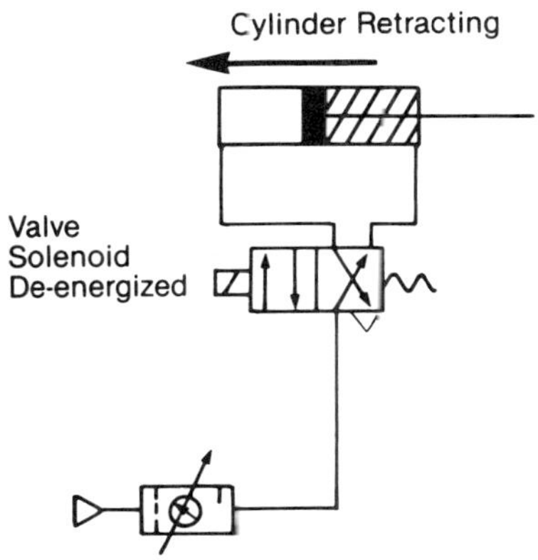

Figure 5.16

5.5. FOUR-WAY, FIVE-PORTED VALVE CIRCUITRY

As pneumatic systems become more complex, greater air valve flexibility is needed. The four-way, five-ported air valve provides that flexibility (Fig. 5.17). The five-ported valve provides the opportunity to use dual pressure, pressure and vacuum, and the option of common individual exhaust.

The four-way, five-ported valve can be used to supply dual pressure to a cylinder (Fig. 5.18). The flow from port 5 to port 4 is at 80 psi, supplying adequate clamping force. Flow from port 3 to port 2, controlled by regulator R2, is at 20 psi, which is enough to return the clamp cylinder. Using the lower pressure to release the clamp conserves energy, lowers operating cost, and extends the life of the equipment.

Figure 5.17 Four-way, five-ported solenoid-operated, spring-return. Two-position valve.

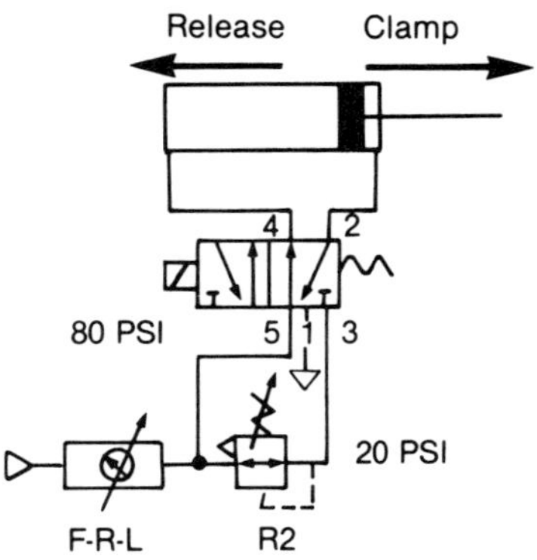

Figure 5.18

The four-way, five-ported valve is also offered in a three-position version (Fig. 5.19). There are three basic center conditions. The first, blocked center, in a typical application traps the air in a cylinder or air motor and restricts motion. This allows a cylinder to support or cushion a load (but will not "hold" a changing load). With the ports blocked, the load on the cylinder causes the piston to move until the trapped air is compressed to

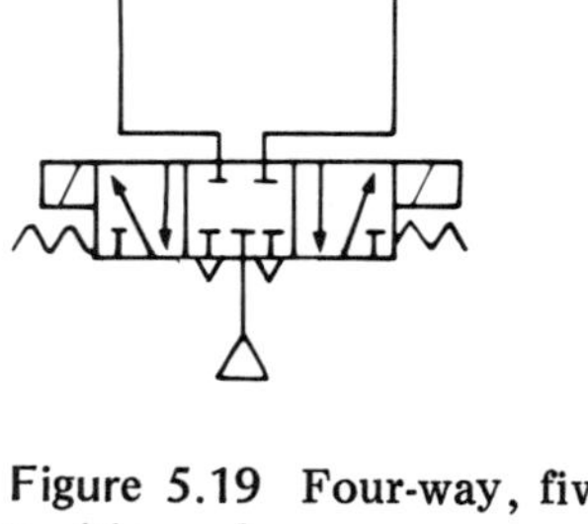

Figure 5.19 Four-way, five-ported double solenoid, spring centered. Three-position valve.

Figure 5.20 Three-position valve. Both cylinder ports connected to pressure in the center position.

a pressure that is high enough to provide enough force to support the load.

The second center position is with the cylinder ports connected to pressure (Fig. 5.20). This condition would be useful for controlling a reversible air motor. When the valve is in the center position, pressure prevents free rotation in either direction and applying pressure to both ports in the fastest way to stop the air motor (apart from an external brake). This would also stop an air cylinder.

The third center position is with cylinder ports open to exhaust (Fig. 5.21). This condition is often used to remove all pressure from the cylinders or air motors so that they can be easily moved during setup or an emergency stop situation.

Cylinder rod speed can be affected by metering pressure into the cylinder or metering exhaust air. Metering exhaust air is the recommended method. When pressure is metered into a cylinder, the load and friction prevent the cylinder from moving until the pressure builds enough to overcome the holding forces. Movement of the cylinder increases the volume of the chamber, causing the pressure to drop, and the cylinder stops until the incoming metered air can repeat the cycle. This results in very erratic cylinder movement.

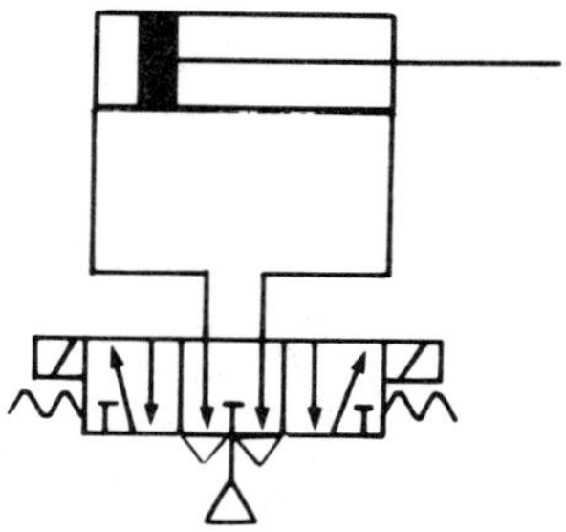

Figure 5.21 Three-position valve. Both cylinder ports connected to exhaust in the center position.

Metered exhaust significantly decreases but does not necessarily remove the problem of erratic cylinder movements. Higher traverse velocities are easier to control by this means because the circuit can respond rapidly to internal or external force changes. Adequate pressure and volume free flowing from the valve creates the necessary force to move the cylinder. The exhaust air is restricted by the flow control, thus limiting the cylinder rod speed by the rate of exhaust flow.

A "sandwich" flow control is shown in Figure 5.22. This flow control is mounted between the valve body and the subbase. This

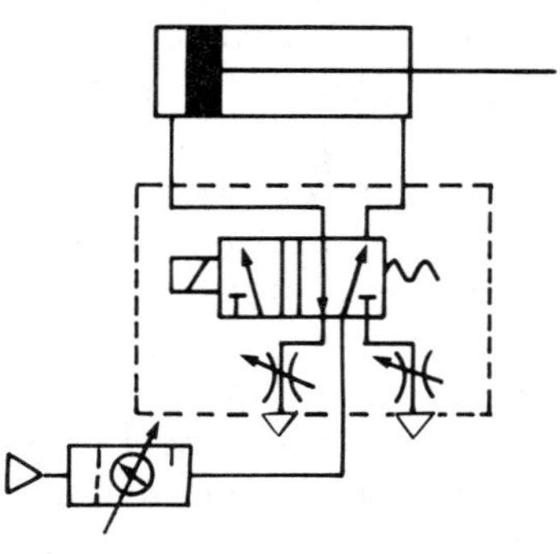

Figure 5.22 Four-way, five-ported valve with sandwich flow control.

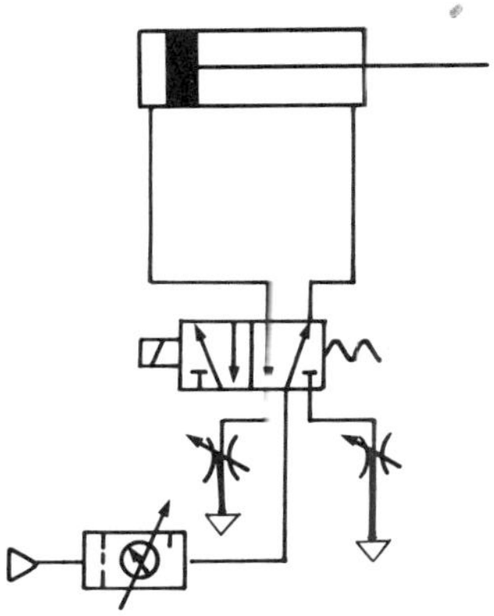

Figure 5.23

configuration provides the space to house an adjusting needle properly sized to the flow of the valve, giving precise metering. The sandwich flow control provides ease of installation and, when a number of valves are used on a manifold, allows metering adjustment of all cylinders from a single location.

An alternative to the sandwich flow control is the exhaust port flow control (Fig. 5.23). This metering device is simply threaded into the exhaust ports on the valve. The performance will be adequate for many applications.

The most accurate control circuit is one in which the flow control is mounted closely coupled at the cylinder port (Fig. 5.24). This is due to the need to bring a smaller volume of exhaust air up to the "controlled pressure" before control can be achieved. The long lines (volume of line), as compared to the cylinder control volume, will greatly affect when the control takes place. Again, higher traverse velocities are easier to control consistently.

A regulator for regulating the air supply to the system is recommended when using any type of metering control. The volume of flow through an orifice varies with pressure. Once a flow control is adjusted (the orifice size determined), fluctuations in system pressure will cause the cylinder rod speed to fluctuate. The greater the thrust force available in relation to the work re-

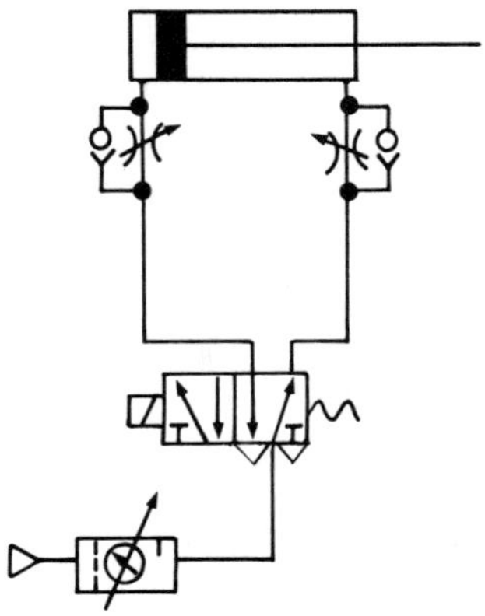

Figure 5.24

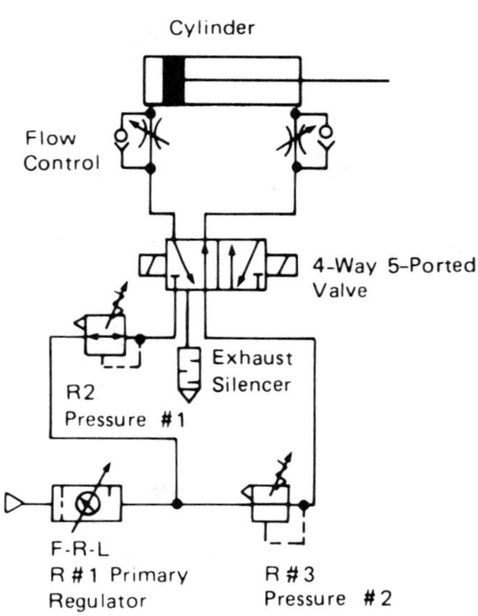

Figure 5.25

quired, the easier it is to control the uniform motion of the cylinder stroke.

The optimum control would use flow control at each cylinder port (Fig. 5.25), a four-way, five-ported valve with dual pressure, and an exhaust silencer for noise abatement. On each machine a primary regulator is recommended to eliminate the fluctuations of the plant air supply. Secondary regulators such as R2 and R3 will then have a constant primary pressure, which provides greater accuracy in maintaining secondary pressure. The regulator spring should be selected to allow the secondary pressure to be within the middle three-fifths of the regulator range. If the required operating pressure is 60 psi, a 0–125 psi spring range should be selected for optimum performance. Secondary pressure should be no higher than 80% or lower than 20% of primary for accurate regulation. But often two speeds are required as a cylinder moves through its stroke. For instance, suppose we have a cylinder with a stroke of 20 in. (508 mm). A fast forward speed of 20 ft/min (0.10 m/s) is needed for the first 10 in. (254 mm) with a speed of 5 ft/min (0.025 m/s) for the remainder of the stroke. The desired return speed is 25 ft/min (0.125 m/s). It is evident that we need three speed controls and at least one directional control valve. By using a five-ported, four-way valve, the fast forward and return speeds can be obtained by installing needle valves in the two independent exhaust ports. The slow speed forward remainder must now be obtained. For this valve capable of sensing the cylinder rod position must be added. A normally open, two-way cam-operated valve can be used. It is placed in parallel with the slow speed flow control so that as the first 10 in. of forward stroke is completed, flow is directed through the newly placed flow control. It may be a difficult task to place the cam in the correct position, because the valve must be actuated soon enough to allow for the buildup of back pressure required for the final traverse velocity. Let us go through the circuit one step at a time.

The de-energized circuit is shown in Figure 5.26. When the main valve is energized, air is directed to the A port of the cylinder (Fig. 5.27). Air exhausing from the B port flows through both cam valve Y and needle valve Z [which is independently set to provide a motion of about 5 ft/min (0.025 m/s)]. Most of the air travels through the open cam valve and then passes through needle

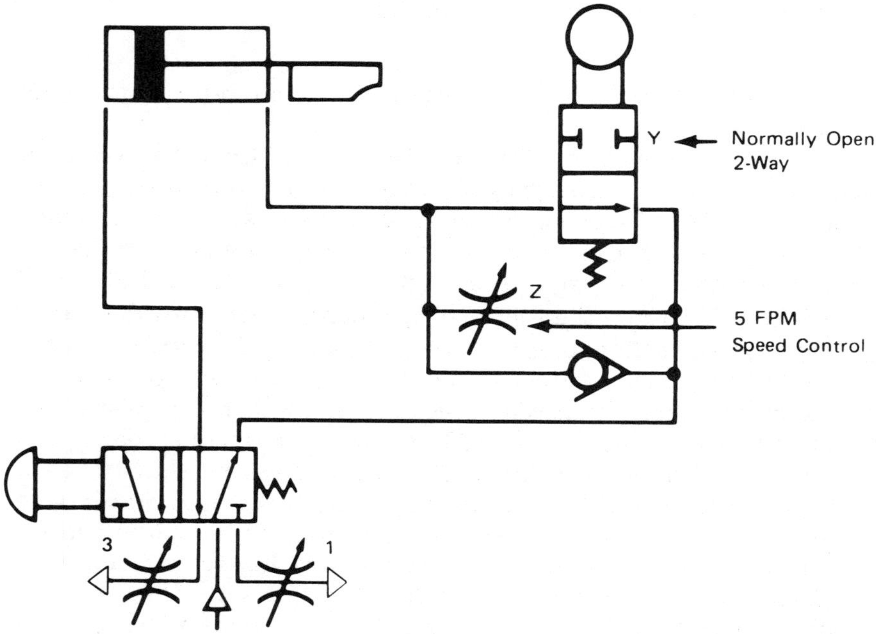

Figure 5.26

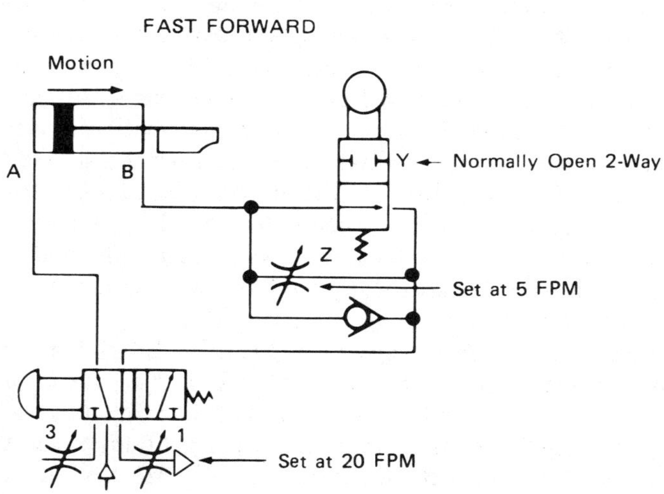

Figure 5.27

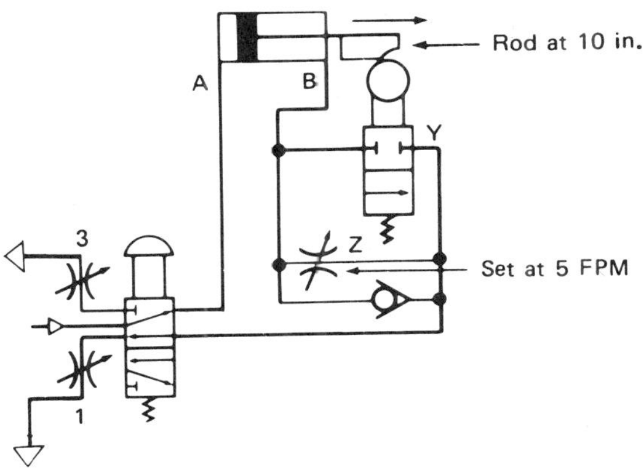

Figure 5.28

valve 1, which is adjusted after Z has been set, so that the two valves exhausting together provide a motion of 20 ft/min (0.10 m/s). When 10 in. (254 mm) of forward stroke is sensed, the cam valve shifts to its closed position (Fig. 5.28). This causes all the air exhausting from the cylinder B port to pass through needle valve Z. Since the valve was independently set to provide motion of about 5 ft/min, the speed of the cylinder is gradually reduced until the new pressure balance is established.

Remember that the exact point of the stroke where this occurs is a variable, depending on the volume in the exhaust lines between the cylinder B port, the line to the cam valve, and the line to the Z needle valve. The entire volume in these lines must be brought to the new equilibrium pressure before the reduced speed can be realized. For retraction (Fig. 5.29), the directional valve is released, thus directing supply air to the cam valve Y, which is closed, needle valve Z, and the check valve. Since the check valve has the least resistance, a nearly unrestricted flow is fed into cylinder port B. Air exhausting from A is passed through the directional control valve and into needle valve 3, which is set at 25 ft/min (0.125 m/s), the required retraction speed.

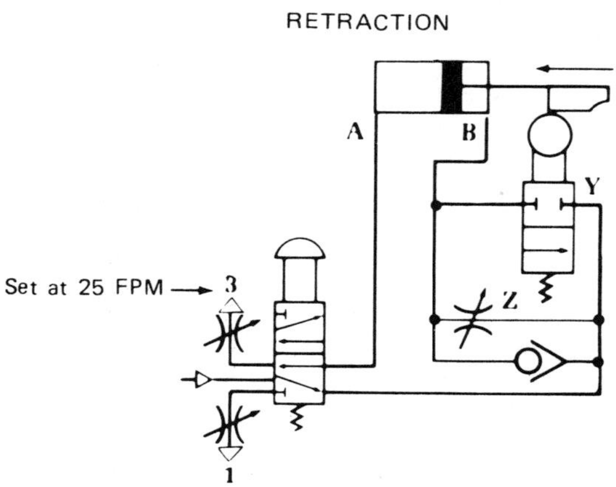

Figure 5.29

Once again, since needle valve 1 is not closely coupled to the cylinder, some problems of control may exist. An improvement is achieved by placing needle valve 1 close to the cylinder (Fig. 5.30). Now, even though adjustments are still interactive, prob-

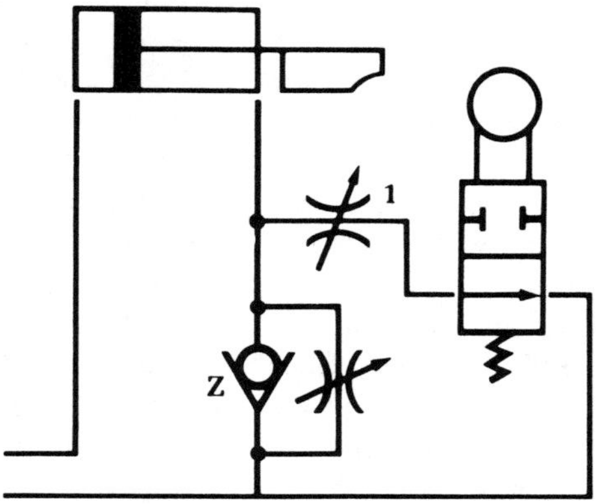

Figure 5.30 Circuit 9-11.

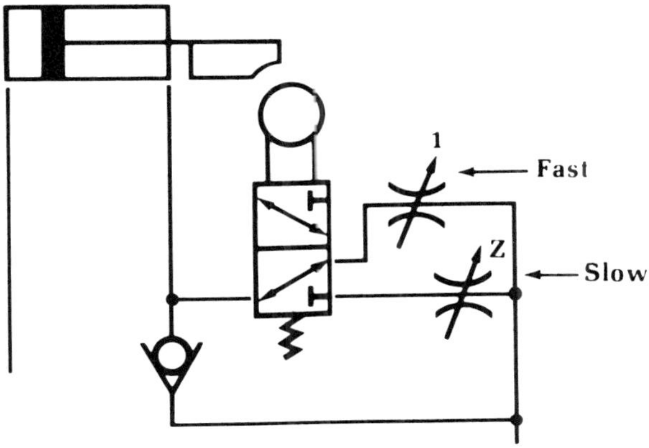

Figure 5.31

lems with control may be diminished because trapped volumes have been reduced.

Another change has been made in Figure 5.31. First, the two-way valve has been replaced by a three-way cam-operated valve. This enables the flow controls to be placed so that minimum interactions take place. Also, they are closer to the cylinder, providing better control.

Finally, Figure 5.32 will give the best control of all the pneumatic circuits shown. The flow controls are even more closely coupled, and the flow control interaction is virtually nonexistent.

Safety note: It is extremely difficult to control the speed of a pneumatic cylinder where the load sensed by the cylinder may vary. This is due to the compressibility of air causing delays in reaching force equilibrium under shifting load conditions. Therefore, each speed control circuit must be examined very closely for all possible failure conditions. Remember that most conventional circuits will not control actuator speeds if the external loads or internal actuator frictions change.

Contrary to most opinions, inching or positioning suitable for many applications can be achieved with a pneumatic cylinder.

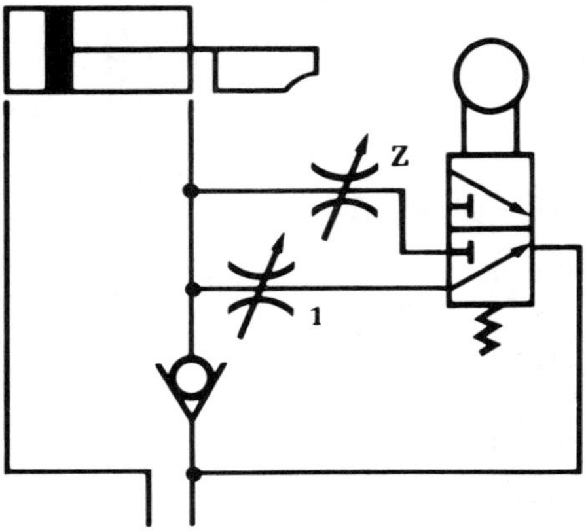

Figure 5.32 Circuit 9-13.

This is done by pressurizing both cylinder ports (Fig. 5.33). The cap end has a larger area exposed to pressure than the rod end due to the rod diameter. When the pressures on both sides of the piston are equal there is greater force on the cap end, causing the cylinder to move toward the rod end. As the cylinder moves,

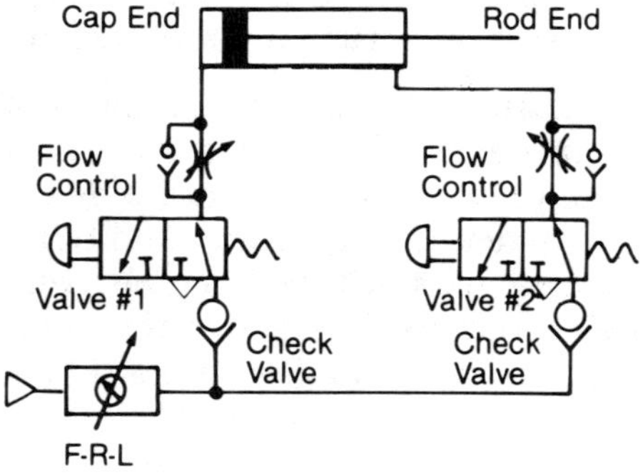

Figure 5.33

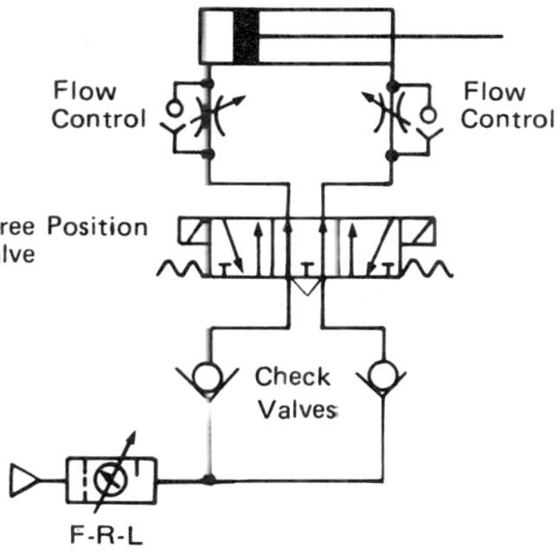

Figure 5.34

check valve 2 seats, causing the piston to compress the air in the rod end. When the pressure increases in the rod end to provide a force equal to the force in the cap end, the cylinder stops. When valve 1 is actuated, pressure on the cap end is vented to exhaust, causing the cylinder to move toward the cap end. When valve 2 is actuated, the cylinder moves toward the rod end.

The same function can be achieved with a three-position valve (Fig. 5.34). The valve must be five-ported to permit isolation of the rod end from the cap end. Flow controls are necessary to prevent erratic movement. Either of these methods will allow a pressure greater than required to move the load to minimize the response time when the valve centers and the cylinder stops.

Another common type of pneumatic circuit is the "sequencing circuit" (Fig. 5.35). This circuit functions without the use of individual power valves for each cylinder. A valve with a C_v of 1.5 will be needed for this circuit. This allows enough flow to operate directly from the lever and mechanically actuated valves in a greater many applications.

In the initial condition, system pressure through valve A and valve C is pressurizing the rod end of cylinder 1 and cylinder 2. When valve A is actuated, the pressure at the rod end of cylinder

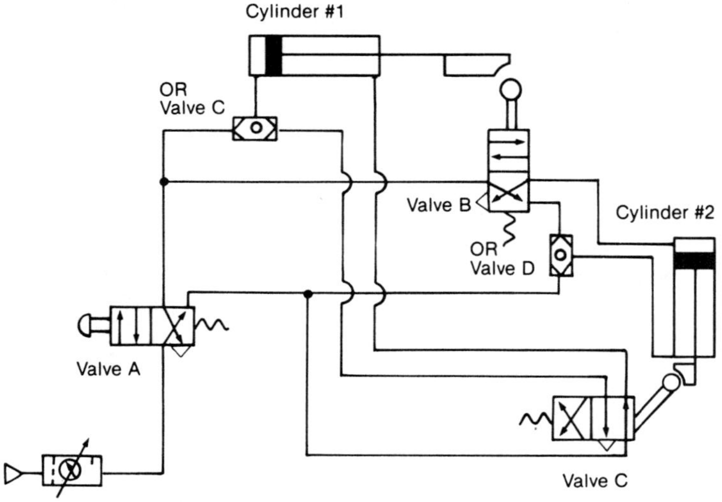

Figure 5.35

1 exhausts. The rod end of cylinder 2 is pressurized through valve B and/or valve D, and the cap end of cylinder 1 is pressurized through "OR valve C." OR valve C could be a quick exhaust valve used as a shuttle valve. This would allow the passage of sufficient air flow. Cylinder 1 then extends actuating valve B. Valve B will exhaust the rod end and pressurize the cap end of cylinder 2. Cylinder 2 will then extend releasing valve C. There is no supply to valve C; therefore, nothing happens.

When valve A is released, the cap end of cylinder 2 is exhausted. Also, the cap end of cylinder 1 is pressurized (maintaining its position) and the rod end of cylinder 2 is pressurized, causing it to retract. When valve C is actuated, the cap end of cylinder 1 will exhaust as the rod end is pressurized. This will retract cylinder 1, completing the cycle.

The use of the shuttle (OR) valves will maintain pressure on both cylinders at all times. This will almost eliminate any erratic movement. The advantages of this circuit are simplicity and low cost.

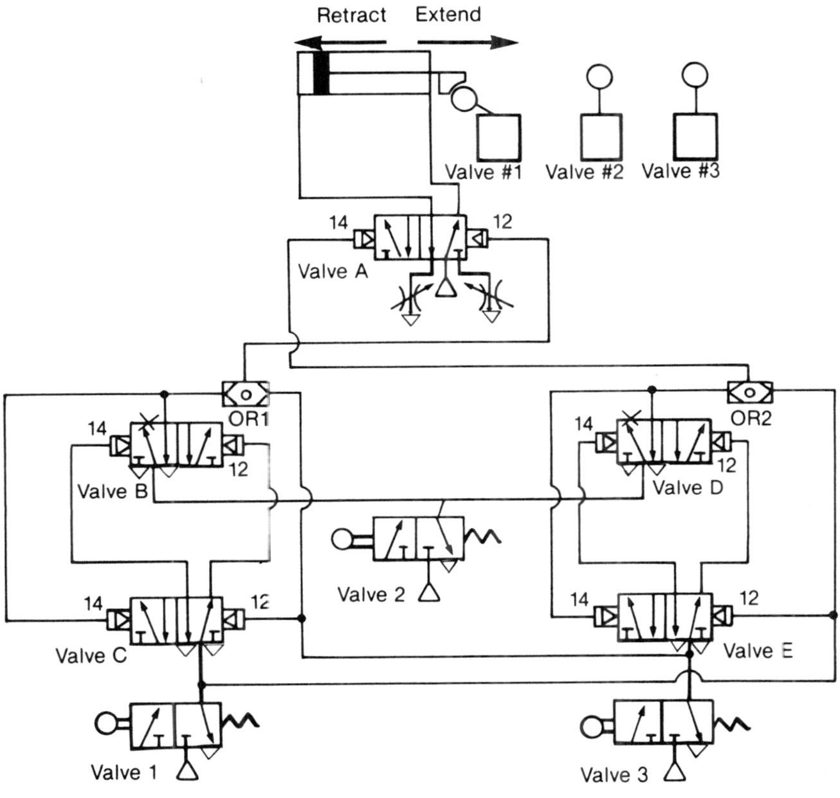

Figure 5.36

Through the use of double air-piloted valves a counting/sequencing type of circuit can be developed (Fig. 5.36). This circuit provides a counting function for valve 2. The first time the cylinder actuates valve 2, the cylinder retracts. The second time valve 2 is actuated, nothing happens. In this case the cylinder simply continues to extension stroke. The third time the cylinder actuates valve 2 the cylinder will extend to the end of its stroke. The fourth time the cylinder actuates valve 2 nothing will happen. The cylinder simply retracts to the initial position.

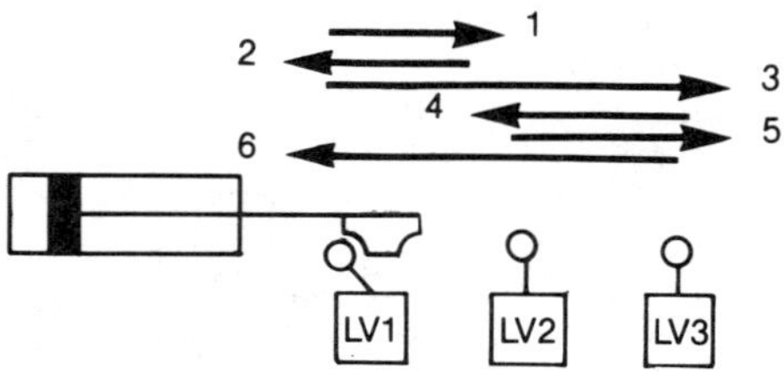

Figure 5.37

The circuit function can also be explained as shown in Figure 5.37. Valves B and D are being used as three-way valves. This is how the circuit operates. The circuit is pressurized in the initial position, which is shown. With the cylinder retracted valve 1 is actuated. Air flows through valve 1, valve C, and actuates the 12 pilot of valve B. Valve 1 also actuates the 12 pilot of valve E and air flows through shuttle 2, actuating port 14 and shifting valve A. Valve 2 is to exhaust. The cylinder now extends. Valve 1 exhausts, relieving pressure on the various pilots. The valves being double-piloted and friction-detented retain their position. The cylinder continues until it makes valve 2. Valve 2 is actuated. However, pressure is blocked by valve D. But air flows through valve B to the 14 pilot port valve C (valve 1 is to exhaust) and through shuttle 1 to the 12 pilot port shifting valve A, causing the cylinder to retract.

The cylinder retracts, exhausting valve 2. The cylinder actuates valve 1, which again pressurizes the 12 pilot port of valves. There is no action at this time because valve 3 is not yet actuated. Valve 1 now lets air flow through valve C to the 14 pilot port on valve B, shifting the valve so that when valve 2 is next actuated the air flow will be blocked. Valve 1 also lets air flow through shuttle 2, shifting valve A and causing the cylinder to extend.

The cylinder now extends, actuating valve 2. Valve 2 is now blocked by valve B and valve D. There is no action and the cylinder continues to valve 3.

Valve 3 is then actuated. Valve 3 pressurizes the 12 pilot port shifting valve C. There is no action; valve 1 is to exhaust. Valve 3 lets air flow through valve E, pressurizing the 12 pilot port and shifting valve D (valve 2 is to exhaust). Valve 3 also lets air flow through shuttle 1, shifting valve A, which retracts the cylinder.

The cylinder retracts and actuates valve 2 for the third time. Valve 2 is now blocked by valve B, but air flows through shuttle 2 shifting valve A, causing the cylinder to extend. Valve 2 also shifts valve E (no action; valve 3 is to exhaust).

The cylinder extends to valve 3 for the second time. Valve 3 will again let air flow through shuttle 1, shifting valve A, causing the cylinder to retract. Valve 3 now lets air flow through valve E to the 14 pilot port of valve D, shifting this valve so that it will block valve 2 when it is next actuated, and valve 3 also shifts valve C, thus preparing for the next cycle.

The cylinder retracts, actuates valve 2 (valve B and valve D again block valve 2; there is no action), and continues on to valve 1, which starts the next cycle.

This circuit can be used to divert packages to different conveyors, to alternate operations, to provide a basis for any type of function that requires a count or alternating-type sequence.

6

Pneumatic Valve Sizing

6.1. INTRODUCTION

Until recently the sizing of pneumatic components, such as directional control valves, filters, and regulators, has been along the lines of convenience rather than efficiency. With the escalating cost of energy circuit designers involved in sizing pneumatic components have found it necessary to give up the practice of rule-of-thumb convenience and are designing more efficient circuits and selecting efficient components to use in those circuits. Many factors influence the "best" circuit design, including component cost, ease of maintenance, space available for mounting components, availability of parts, energy cost, and so forth. The best circuit design or system is in reality a compromise between all these factors and more. In an effort to obtain the best compromise in the area of pneumatic directional control valve sizing, a method for sizing a value for a given set of conditions has been developed and has gained wide acceptance. In this method the directional control valve is sized by utilizing its flow coefficient.

6.2. FLOW COEFFICIENT C_v

When selecting pneumatic directional control valves, an important consideration is their ability to pass the required volume of air at an acceptable pressure drop. The pressure drop across a component, measured in a conventional way, is sometimes used as a measure of the flow rating of that component. A common method of rating flow is by a C_v factor.

The C_v factor is derived from an expression which gives the number of gallons of water per minute that will pass through the valve with a 1-psi differential between the valve's inlet and outlet. It is calculated from experimental data with a formula that will be described later.

Typical capacity coefficients for four-way directional valves are:

Port pipe size (national pipe thread)	C_v range
1/8 in.	0.2 to 0.8
1/4 in.	0.5 to 2.0
3/8 in.	1.5 to 3.5
1/2 in.	2.5 to 5.0
3/4 in.	4.0 to 9.0
1 in.	6.0 to 16.0

A rough rule for estimating attainable flow values of four-way, manifold-mounted valves was suggested by one of the authors in the early 1960s.* The C_v of a "large" flow capacity four-way valve, subbase-mounted, can be estimated by the square of the flow port size expressed in quarter-inches. Thus, a 1/2-in. valve is two quarter-inches and $2^2 = 4$, the expected approximate C_v. Similarly, a 1/8-in. valve is a half of a quarter-inch and $(1/2)^2 = 0.25$, the expected flow factor for this valve.

In many valve designs, the capacity between different flow paths may vary up to 50%. One manufacturer's 1/2-in. port valve may actually pass less flow at a given pressure drop than another 1/4-in. port valve. Thus, the designer would do well to specify valves that have at least the typical flow coefficients shown.

6.3. SIZING A VALVE FOR FLOW (FOR U.S. UNITS)

The capacity coefficient formula is:

$$C_v = \frac{Q}{22.48} \sqrt{\frac{T_1 G}{\Delta P(P_2 + P_a)}}$$

where:

C_v = capacity coefficient (a number)
Q = flow in standard cubic feet per minute (scfm) at 14.7 psig and 36% relative humidity
G = specific gravity of the flowing medium ($G = 1$ for air)
T_1 = absolute temperature (460 + degrees Fahrenheit) of inlet air
P_1 = inlet pressure (psig)
P_2 = outlet pressure (psig) ($P_2 = P_1 - P$)
ΔP = pressure drop (psi), static inlet to static outlet pressure
P_a = atmospheric pressure (psia)

Notes:

1. The use of the C_v formula is accurate to perhaps ±10% for pressure drops of 2 psi or less. Larger pressure drops will result in larger errors. For a more accurate estimation of

---

**Proceedings—1962 International Fluid Power Conference.*

the actual flows that may be attained, use the flow curves published by the valve manufacturer for the specific valve you plan to use.

2. This equation is valid for subsonic flow only. To ensure subsonic flow (flow velocities below the speed of sound, 1100 ft/s), limit pressure drop so that

$$\frac{P_2 + P_a}{P_1 + P_a}$$

is between 0.85 and 1.0.

3. The effect of relative humidity (RH) on G for air is 0.6% over the range 0 to 100% RH and may be ignored for ordinary applications.

Although this formula is not difficult to work with, we can make use of tables for some of the basic quantities found in the equation. It may be rewritten as follows for U.S. units:

$$C_v = \text{flow rate (cfm)} \times A$$

where flow rate is the standard cubic feet of free air per minute (scfm) and A is a constant (Fig. 6.1) which is a function of inlet pressure and pressure drop.

The air-flow rate for a particular rotary system is an easy parameter to obtain because air tools are typically rated in terms of so many cubic feet per minute at a particular pressure. For cylinder application, the number of cubic feet per minute must be calculated.

The A factor is a variable that is a function of two parameters, inlet pressure (psig) and pressure drop through the interior of the valve. Three A constants are given for one primary pressure. Pressure drop is given for three pressure drop values. Typically the A constant at a 5-psi (0.34-bar) drop is used for most applications. In very critical applications where maximum efficiency is needed, a maximum drop of 2 psi (0.14 bar) should be used. In some cases, a 10-psi (0.69-bar) pressure drop is not detrimental and can save money and space. Let us consider some examples.

Example 6.1 An impact tool is being used in a particular circuit. The tool uses 10 cfm (4.72 dm^3/s) at 90 psi (6.2 bar). What is the valve size (C_v rating) and what type of valve is needed?

Solution Since the system needed will be used to power an impact tool, a simple two-way valve is sufficient. The size of the valve is determined by its C_v:

$$C_v = \text{cfm} \times A$$

In this application, we have decided to allow a 10-psi drop across the valve to save on initial investment. This means that the upstream pressure of the valve is:

$$\begin{aligned} \text{Upstream pressure} &= \text{impact tool pressure} + \text{pressure drop} \\ &= 90 + 10 = 100 \text{ psi (6.9 bar)} \end{aligned}$$

The A factor from the chart for 100-psi inlet pressure is 0.031. The C_v will be:

$$\begin{aligned} C_v &= 10 \times 0.031 \\ &= 0.31 \end{aligned}$$

Let us now work a valve sizing problem with a cylinder incorporated into the circuit. Note that the example is only an approximation because when the valve is actuated, air flows into the empty volume of line connecting the valve to the cylinder and the initially empty cylinder volume. As these volumes are pressurized, at some point the pressure rises sufficiently to overcome the load and static friction to start moving the cylinder and its connected load.

Example 6.2 A cylinder with a 6-in. (152-mm) bore and a 25-in. (635-mm) stroke is extended at a pressure of 50 psi (3.45 bar). It must reach the end of its stroke in the relatively long time of 5 s. What size valve should be used? Assume that the valve is mounted as closely as possible to the cylinder and that the connecting lines are consistent with the valve size.

Solution First of all, the average flow rate (cfm) must be found.

Inlet Pressure (psig)	Compression Factor	"A" Constants for Various Pressure Drop		
		2 psi Δ P	5 psi Δ P	10 psi Δ P
10	1.6	.155	.102	
20	2.3	.129	.083	.066
30	3.0	.113	.072	.055
40	3.7	.097	.064	.048
50	4.4	.091	.059	.043
60	5.1	.084	.054	.040
70	5.7	.079	.050	.037
80	6.4	.075	.048	.035
90	7.1	.071	.045	.033
100	7.8	.068	.043	.031
110	8.5	.065	.041	.030
120	9.2	.062	.039	.029
130	9.9	.060	.038	.028
140	10.6	.058	.037	.027
150	11.2	.056	.036	.026
160	11.9	.055	.035	.026
170	12.6	.054	.034	.025
180	13.3	.052	.033	.024
190	14.0	.051	.032	.024
200	14.7	.050	.031	.023

Figure 6.1 Compression factors and "A" constants.

$$\text{cfm} = \frac{\text{volume (in.}^3) \times \text{ compression ratio}}{\text{stroke time (s)} \times 28.8}$$

The volume of the cylinder may be calculated first.

$$\begin{aligned} V\ (\text{in.}^3) &= \text{piston area(in.}^2) \times \text{stroke (in.)} \\ &= 28.26 \times 25 \\ &= 706.5 \text{ in.}^3\ (11.5 \text{ dm}^3) \end{aligned}$$

The compression ratio can be found in Figure 6.1 and is 4.4. The average flow rate becomes

$$\text{cfm} = \frac{\text{volume (in.}^3) \times \text{ compression ratio}}{\text{stroke time (s)} \times 28.8}$$

$$= \frac{706.5 \times 4.4}{5 \times 28.8}$$

$$= 21.6 \text{ cfm } (10.2 \text{ dm}^3/\text{s})$$

The C_v can now be calculated. Let us consider that a highly efficient circuit is needed, therefore, a maximum 2-psi drop is suggested. This means that the A factor is about 0.091.

$$C_v = \text{cfm} \times A$$

$$= 21.6 \times 0.91$$

$$= 1.96$$

6.4. SIZING A VALVE FOR FLOW (FOR SI UNITS)

Another way to size a valve for flow is by calculating the C_v. The equation for C_v is:

$$C_v = \frac{Q}{114.4} \sqrt{\frac{T_1 G}{\Delta P(P_2 + P_a)}}$$

where:

C_v = capacity coefficient (a number)
Q = flow in standard cubic feet per minute (dm^3/s) at 7600 mm Hg, 20° C, and 36% relative humidity
G = specific gravity of flowing medium (G = 1 for air)
T_1 = absolute temperature (273 + centigrade degrees)
P_1 = inlet pressure (bars)
P_2 = outlet pressure (bars)
ΔP = pressure drop (bars) static to static pressure
P_a = absolute pressure

Notes:

1. The effect of relative humidity on G for air is 0.6% over the range 0 to 100% RH and may be ignored for ordinary circuit calculations.

2. This equation is valid for subsonic flow only. To ensure subsonic flow, limit pressure drop so that

$$\frac{P_2 + P_a}{P_1 + P_a}$$

is between 0.85 and 1.0.

In some applications, where the stroke time is short, we must check another valve parameter, that of valve response time, and also consider the effect of the volume of connecting lines which must also be pressurized and exhausted.

6.5. VALVE RESPONSE TIME

Many manufacturers provide data on the time required to pressurize specific volumes. A typical set of data includes, with 100-psig supply, the time required to fill from 0 to 90 psig and exhaust from 100 to 10 psig. This is measured from the instant of energizing, or de-energizing with a solenoid working at rated voltage and frequency. Some typical performance data for relatively fast pilot-operated valves are given in Table 6.1.

Table 6.2 shows the effect on the shift time of a two-way direct solenoid-operated valve with changing outlet pressure and then the effect of changing voltage (power) to the solenoid.

The response time of control valves is becoming more important as pneumatic circuitry is used to control rapidly moving machinery or to provide split-second accuracy of operation. A valve's operating time is important in applications where a command or signal given to the valve requires almost instantaneous (and consistent) response of the device controlled by the valve. Generally speaking, the shorter the response time of a valve, the more reliable it will be in its actual application in a high-speed or high-cycle-rate circuit. The matching of a valve to a specific response needed should take into consideration the total volume being pressurized (and exhausted) (Fig. 6.2 and 6.3) and the time available for the actuation of the valve itself.

Where small volumes are to be pressurized, small ported valves may give a faster response than large ported valves. This is because large valves must pressurize sizable internal volumes as well as the

Table 6.1 Time (seconds) to Pressurize or Exhaust Fixed Volumes*

Port pipe size (NPT)	12 in.^{3}	100 in.^{3}	1000 in.^{3}
1/4 in.	0.060	0.300	—
3/8 in.	0.040	0.090	—
1/2 in.	0.035	0.075	—
3/4 in.	0.030	0.070	0.55
1 in.	—	0.065	0.40
1 1/4 in.	—	0.060	0.37

*Times are measured from instant of energizing or de-energizing a 115 V, 60 cps AC solenoid. Supply pressure is 100 psig. Volumes are filled from 0 to 90 psig, or exhausted from 100 to 10 psig.

Table 6.2 Shift Time: Direct Solenoid-Operated Two-Way Valves

$\frac{\text{Outlet pressure}}{\text{Inlet pressure}}$	DC voltage	Shift time (ms)	
		Gas	Water
1.00	12	4.25	6.75
0.83	12	4.75	7.25
0.67	12	5.25	7.75
0.50	12	5.50	8.25
0.33	12	6.00	8.75
0.19	12	6.25	—
0.10	12	—	8.75
0.67	13.0	4.75	7.50
0.67	12.0	5.25	7.75
0.67	11.5	5.50	7.75
0.67	11.0	5.50	8.50
0.67	10.0	6.25	9.25

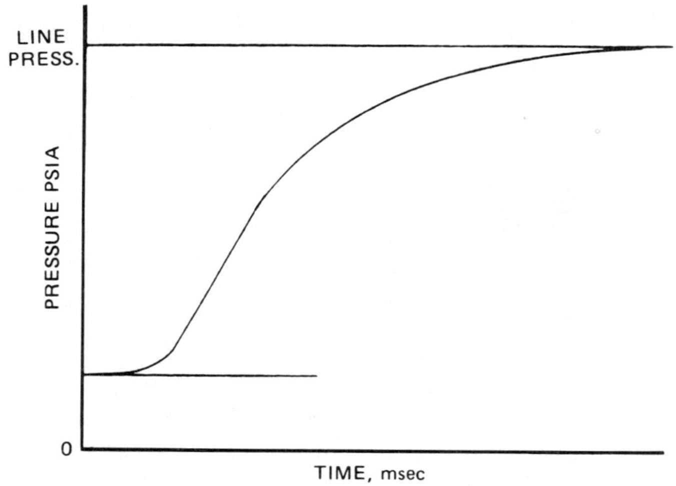

Figure 6.2 Typical curve for filling large chambers.

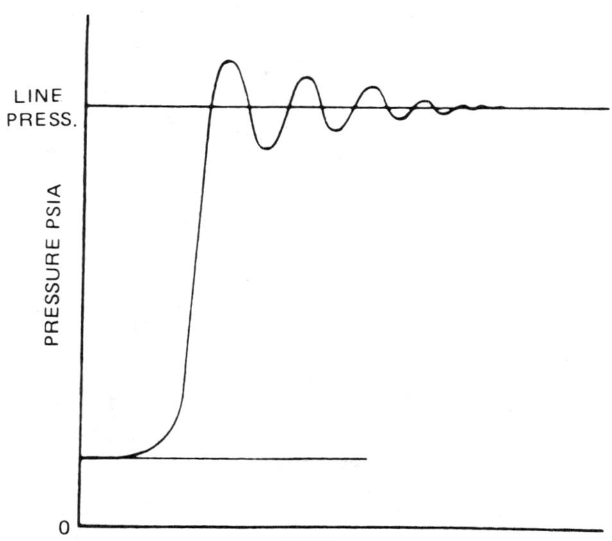

Figure 6.3 Typical curve for filling small chambers.

loads connected to them. Thus, even though they have large flow capacities, the time required to pressurize them may make large valves appear to have a slow response when used with relatively small loads.

An additional delay and timing variable results from the variation of the sinusoidal electric voltage at the instant the signal voltage is switched on (or off) to the solenoid. This variation amounts to 8.3 ms or more, depending on the solenoid's electrical inertia.

Concern for correct component sizing should not only be concentrated on the pneumatic directional control valve, but should encompass the entire system. That is, filters, regulators, and lubricators must also be sized for best system performance. At this point let us briefly discuss the sizing of each of these pneumatic system components.

6.6. SELECTING A FILTER

Let us now select a filter. When selecting a filter, these three steps should be followed:

1. Determine the maximum flow requirements at which the filter will be used. This will be in terms of cubic feet per minute (dm^3/s).
2. Determine the allowable pressure drop at the needed flow.
3. Refer to the filter manufacturer's flow curves on the filter to find one that offers the particular flow rate at the predetermined pressure drop.

Remember, if an element is sized too small, the result will typically be a high initial pressure drop at the rated flow. Also, the element may "load up" quickly at the start of operations with a resulting increasing pressure drop. This can lead to possible air starvation downstream. When the pressure drop reaches 10 psi, the filter element should be changed or cleaned in order to prevent high pressure losses. As most pneumatic filters are designed and constructed so that air is admitted into the bowl with a pronounced swirling action, the centrifugal forces resulting from the

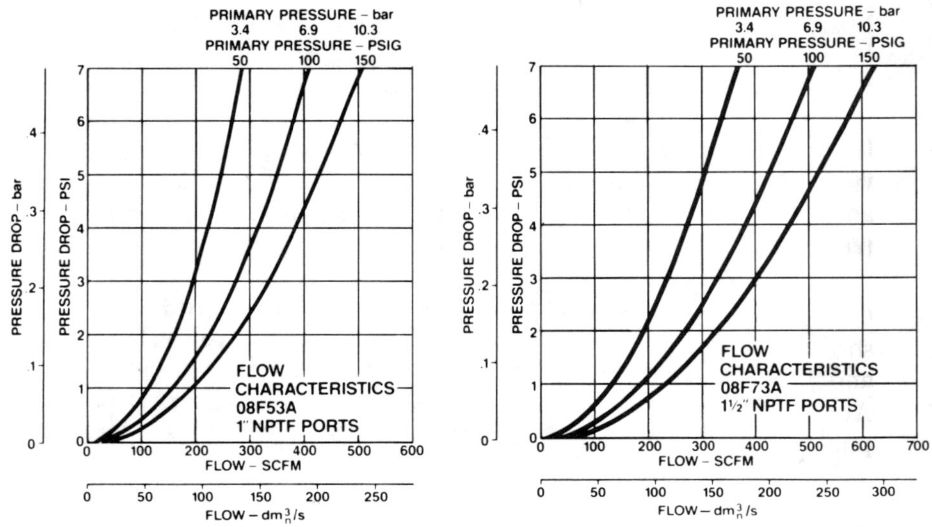

Figure 6.4

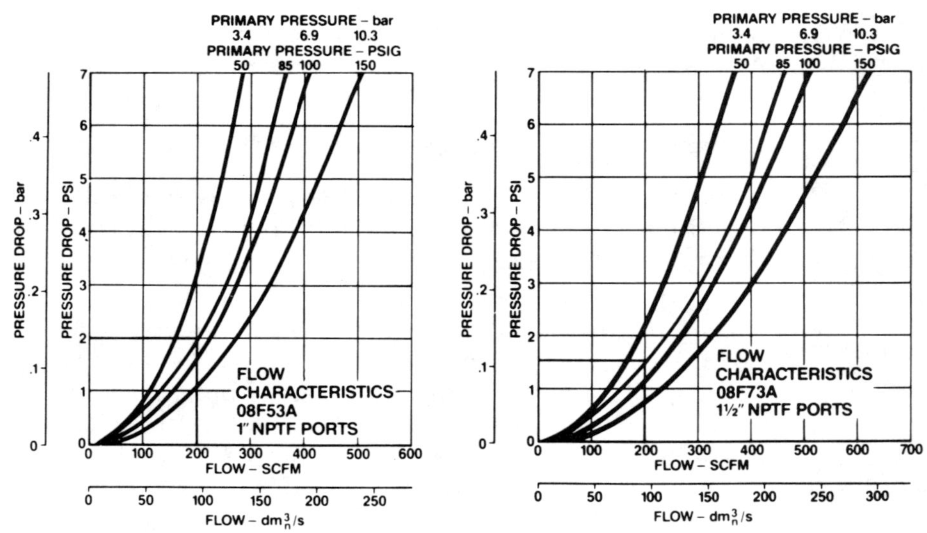

Figure 6.5

swirl throw the entrained solids out against the inner surface of the filter bowl. The air then enters and passes through the filter element with very little contaminant to be removed. Of course, if the filter bowl fills with water, dirt is forced into the element quite rapidly, sometimes with devastating results. This can be avoided by frequent draining or by using an automatic bowl drain device. On the other hand, if the filter housing is too large, the low swirl velocity generated may result in poor contaminant and condensate separation.

Example 6.3 The XYZ Machinery Company is using a radial-type piston motor which develops a maximum output of 10 hp (7.5 kW) at 800 rpm. Its air consumption, as read directly from catalog data, is 200 cfm (95 dm^3/s). Select a filter that will work with a maximum pressure drop of 2 psi (0.14 bar). The data given for the motor are at 85 psi (6 bars).

Solution From motor catalog data it is evident that we need a filter that will pass 200 cfm at 85 psi, with a 2-psi pressure drop. As we go through a filter catalog we come upon the two graphs in Fig. 6.4, which show the flow characteristics of a basic 1-in. port body with 1-in. and 1 1/2-in. pipe ports. Since the primary pressure of 85 psi (6 bars) is not given, it must be estimated as shown in Fig. 6.5.

Working with the 08F53A graph, we find a 2-psi (0.14-bar) drop for 200 cfm if 1-in. ports are used. A 1.5-psi (0.1-bar) drop is evident for the 08F73A if 1 to 1 1/2-in. ports are used. Either may be selected, but the former has better water removal efficiency.

Example 6.4 A 6-in (152-mm) bore, 3-in. (76.2-mm) rod, 10-in. (254-mm) stroke cylinder extends in 4 s and retracts in 3 s. The working pressure required at the cylinder is 50 psi (3.45 bars). The filter's primary pressure is 100 psi (6.9 bars). What filter is needed?

Solution The flow rate must be calculated first:

$$\text{cfm} = \frac{\text{volume (in.}^3\text{)} \times \text{compression ratio}}{\text{time to fill cylinder} \times 28.8}$$

For extension:

$$\text{cfm} = \frac{(28.27 \times 10)[(50 + 14.7)/14.7]}{4 \times 28.8}$$

$$= 10.8 \text{ (or } 5.09 \text{ dm}^3/\text{s)}$$

For retraction:

$$\text{cfm} = \frac{(\pi/4)(6^2 - 3^2) \times 10 \times [(50 + 14.7)/14.7]}{3 \times 28.8}$$

$$= 10.8 \text{ (or } 5.09 \text{ dm}^3/\text{s)}$$

The filter must be able to pass 10.8 cfm (or 5.09 dm^3/s). For either direction of cylinder motion, with a primary pressure of 100 psi (6.9 bars), the allowable pressure drop is 1.5 psi (0.1 bar). Looking through the catalog, we find the graphs shown in Fig. 6.6.

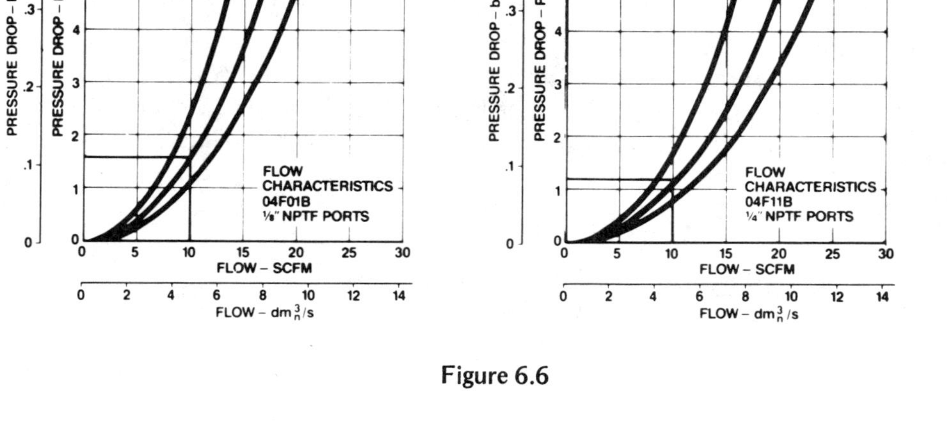

Figure 6.6

The filters selected will be the 04F11B type with 1/4-in. ports, which will deliver 10.8 scfm (5 dm^3/s) with a 1.2-psi (0.08-bar) pressure drop.

However, it is good practice to select a large filter to facilitate ease of piping. A 6-in. bore cylinder typically would have 1-in. ports. If this larger filter is selected than it is essential that the component's contaminant removal efficiency be examined.

The next component that requires careful attention is the air line regulator.

6.7. SIZING A REGULATOR

A regulator should be mounted as close as possible to the actuator it serves and sized in such a way that it provides the needed flow rate with a relatively low pressure drop at the desired pressure rating.

The equation for flow rate for a cylinder has been given. For U.S. units:

$$\text{cfm} = \frac{\text{volume (in.}^3) \times \text{ compression ratio}}{\text{time to fill cylinder(s)} \times 28.8}$$

For SI units:

$$\text{Flow rate (dm}^3/\text{s)} = \frac{V\ (\text{dm}^3) \times \text{ compression ratio}}{\text{time to fill cylinder (s)}}$$

Once flow rates are found, a regulator may be selected. Pressure drop, reverse flow capacity, venting, and regulation characteristics may then be checked.

Example 6.5 A cylinder with a 4-in. (101-mm) bore, 2-in. (51-mm) rod, and 20-in. (508-mm) stroke, extends in 2 s and retracts in 1 s. The pressure for extension is 60 psi (4.14 bars) and for retraction is 20 psi (1.38 bars). What size regulator should be used for this circuit?

Solution Since the regulator is controlling the pressure of the air supplied to the directional control valve, both extension and retraction must be considered.

Flow rate for extension:

$$\text{cfm} = \frac{V \times \text{compression ratio}}{\text{time} \times 28.8}$$

$$= \frac{12.56 \times 20 \times [(60 + 14.7)/14.7]}{2 \times 28.8}$$

$$= 22.2 \text{ (or } 10.5 \text{ dm}^3/\text{s)}$$

Flow rate for retraction:

$$\text{cfm} = \frac{\pi [4^2 - 2^2)/4] \times 20 \times [(20 + 14.7)/14.7]}{1 \times 28.8}$$

$$= 15.4 \text{ (or } 7.3 \text{ dm}^3/\text{s)}$$

The regulator will be sized for at least 22.2 cfm. If a basic 1/4-in. ported body with 1/4-in. ports is selected, the characteristics shown in Figure 6.7 can be found from catalog data of most manufacturers of air line pressure regulators. Figure 6.7 shows the relief and flow characteristics of the regulator. From this graph it can be noted that a 5-psi (0.34-bar) decrease will occur from the

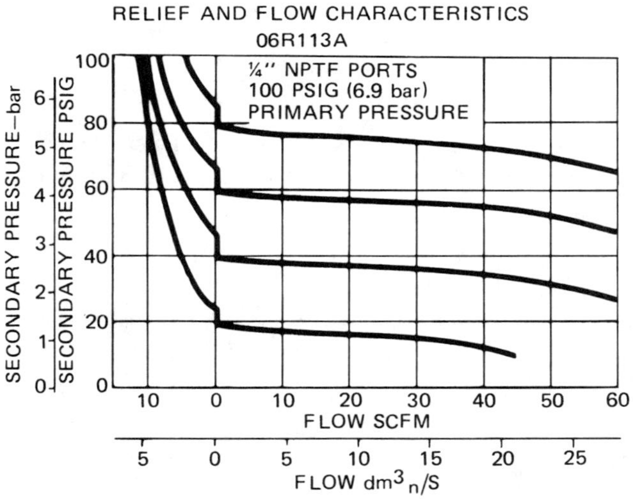

Figure 6.7 Relief and flow characteristics of regulators.

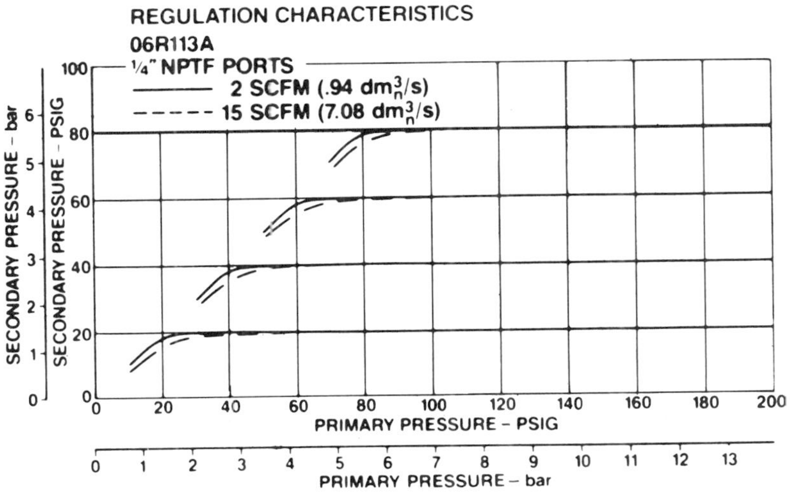

Figure 6.8 Regulation characteristics.

set point at a flow of 22 cfm. Also, venting characteristics can be found at the left of the 0 cfm line.

Figure 6.8 shows the regulation characteristics of the regulator. It is sometimes referred to as the supply pressure effect. In other words, how much will the secondary (output) pressure vary with changes in primary pressure? This regulator exhibits excellent characteristics, as noted in the graph.

One of the most often incorrectly sized and improperly adjusted components is the air line mist-type lubricator.

6.8. SIZING A LUBRICATOR

The lubricator, like the filter, is sized on the basis of air flow rate and must have sufficient air flow to achieve its intended pressure difference (drop) in order to function. Once again, the maximum flow requirements of the system must be determined and the pressure drop [usually in the range of 1 to 5 psi (0.07 to 0.34 bar)] for various body sizes and plumbing must be determined.

Then catalog data are checked and a lubricator is selected.

Note: OSHA regulations (29 CFR 1910.1000) place a limit of 5 mg/m^3 of oil mist in air. This is approximately equivalent to 3.5 fluid ounces (91 grams) of oil in the total volume of a plant area of about 40,000 to 50,000 ft^2. This suggests that oil reclassifier devices may need to be installed on exhausts of subsystems.

Example 6.6 A die grinder employs a vane air motor. At maximum output the motor needs 16.3 cfm (7.7 dm^3/s) at 50 psi (3.45 bars). What lubricator should be used?

Solution The lubricator selected must be able to pass 16.3 cfm at a reasonable pressure drop. The pressure drop selected will be 5 psig (0.34 bar). The graphs shown in Fig. 6.9 are those typically found in a catalog for "micromist" lubricators.

The pressure drops for 1/2-in. body size lubricators are 2.3, 1.5, and 1.3 psi for 1/4-, 3/8-, and 1/2-in. ported, respectively. Most lubricators must have a pressure drop to pick up lubricant and atomize it. The pressure drop to provide this generally falls in the range 1 to 5 psi (0.07 to 0.34 bar). Since the plant engineer usually prefers to buy a lubricator with the same size ports as the motor, the 1/4-in. model is selected.

The point has now been reached where the designer can turn his or her attention from the sizing of individual components to examining the overall system performance.

In order to accomplish this the designer must know the principles of application and the performance characteristics of the components being used and also the factors affecting their performance under actual working conditions.

The designer must keep in mind that each component, as well as the overall system, must be properly sized to allow the required mass rate of air to flow at an acceptable pressure drop. It should be remembered that a single incorrectly sized component in the system could cause pressure drop equal to 100 ft or more of piping.

Because the majority of industrial pneumatic systems operate at a compressor pressure of only 100-125 psi, pressure loss or pressure drop is of great concern to the energy-conscious designer. Therefore, it it good engineering practice to estimate or measure

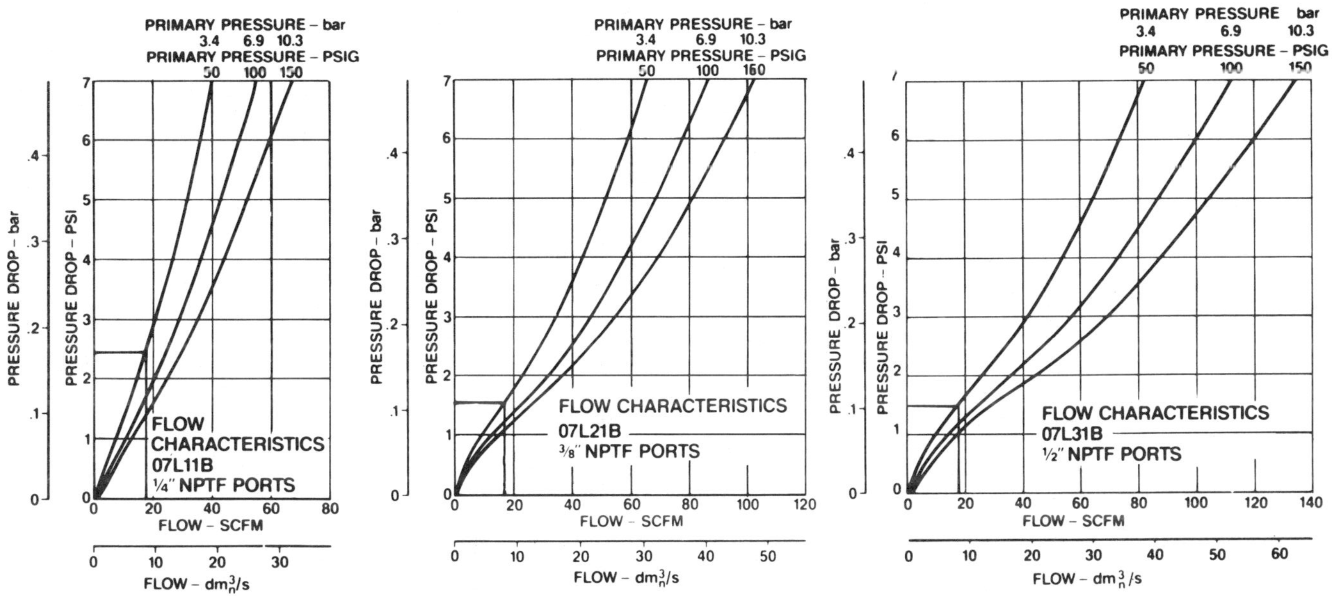

Figure 6.9 Performance characteristics of lubricators.

the system pressure loss, to ensure that it is a minimum or in the acceptable range.

6.9. PNEUMATIC SYSTEM PRESSURE DROP*

Today there are many methods in use for estimating the pressure drop in a pneumatic system. These methods range from the complex to the overly simplified "rule of thumb." However, a simplified technique has been developed to assist the designer who wants to eliminate the confusion present in some methods.

The foundation for this technique is given in the flow factor equation:

$$\text{Flow in standard cfm} = F\sqrt{4/3}\,\sqrt{P_u{}^2 - P_d{}^2} \tag{6.1}$$

However, this and other similar equations that relate the F factors for various components in series are considered complex, difficult, and time-consuming by nontechnical applications personnel. This is very discouraging to the designer, who infrequently encounters air system problems unless he or she programs a programmable calculator or small computer to help solve the problem. By incorporating a special M factor, a set of labor-saving nomographs and charts has been developed. This graphic approach has reduced much of the complexity involved.

As an example, components connected in series can be represented by an F factor equation. By normalizing the above equation all but the individual flow factors are eliminated. Then the following would be a combined flow factor for three components connected in series:

$$\text{System } F \text{ factor} = \frac{F_1F_2F_3}{\sqrt{(F_1F_2)^2 + (F_2F_3)^2 + (F_3F_1)^2}} \tag{6.2}$$

This equation can be further simplified to represent a given length of pipes:

*The material for Sects. 6.9, 6.10, and 6.11 is based on work by Gil Riske, Senior Engineer, Pneumatic Division, Parker Hannifin Corporation; see Appendixes A and B. Material also taken from the following article by Gil Riske; Pressure losses in pneumatic systems. *Product Engineering 30*(44), October 26, 1959.

$$\text{System } F \text{ factor} = \frac{F_r}{\sqrt{M_1 + M_2 + M_3}} \tag{6.3}$$

where F_r is a convenient reference flow factor such as F_1, the flow factor for the first component. Also, the M factors would be

$$(F_r/F_1)^2, (F_r/F_2)^2, \text{ and } (F_r/F_3)^2$$

for M_1, M_2, and M_3, respectively.

Utilizing this simplified equation, it now becomes possible to take the flow restrictions of valves, hose, fittings, etc., and change them into an equivalent length of pipe. Also, an equivalent length of pipe could be broken up into various flow restrictions. Thus, a series system can be treated as a single restriction that follows the above equation. The same equations can be used when designing a pneumatic system.

6.10. PNEUMATIC SYSTEM DESIGN

The vast majority of pneumatic system design problems can be solved by following three basic steps:

1. Utilizing a standard unit of measurement, determine the flow resistance of each component and add the resistances together.
2. Beginning with the air work device and working in a direction that is opposite to the air flow, determine the minimum pressures in each section of the system.
3. Start at the air supply and determine the actual pressures in each section while working toward the end of the system. Work in the direction of the flow.

To simplify the pressure loss calculations in step 1, substitute equivalent lengths of pipe for all restrictions such as fittings and components. If the fittings and components manufacturers do not list equivalent lengths of pipe in their literature, then simple tests will be required and will yield a reliable number.

In order to simplify and shorten the time required for step 1, several charts and nomographs have been developed. These will aid in converting flow path into equivalent lengths of pipe.

If the flow factor F is known, it can be substituted directly into the nomograph in Figure 6.10 and converted into an equivalent length of pipe. This will permit equal-diameter lengths of pipe to be added together. The nomograph in Figure 6.10 relates flow factor F, equivalent length of pipe, and pressure loss.

To determine an equivalent length of pipe, scales A, C, E, and F must be used. To begin, a straight line must be drawn from the flow factor in column A to the nominal schedule 40 pipe diameter in column E. The flow factor for 1 ft of pipe is read from column F. Also, an equivalent length of pipe is given in column C.

The same scales (A, C, E, and F) can also be used as a conversion for pipe sizes. As an example, convert an equivalent length of 3/4-in. schedule 40 pipe to 3/8-in. schedule 40 pipe with the same flow factor F. Using column A as a pivot point, move the right-hand end of the straight line vertically up column E from 3/4 to 3/8 in. Now, the new equivalent length can be read in column C.

However, the inside diameters of the other pipe schedules are different. If one of these other pipe schedules is used, enter Figure 6.11 with the actual pipe inside diameter. This will directly yield a flow factor, which can be substituted into column F of the nomograph.

To determine the pressure loss, scales A, B, and D should be used. Begin with the correct flow factor in column A and draw a line from this point through the expected flow in column B. Extend the line to column D. The value read in this column represents $\sqrt{P_u^2 - P_d^2}$. A quick and easy solution for P_u and P_d can be obtained by using Figure 6.12.

The nomograph can be adjusted for pipe with diameters larger than 2 1/2 in. Columns A, B, F, and E must be modified. All values in columns A, B, and F must be multiplied by 100. The new pipe diameters in column E should be filled next to column F as follows:

2 1/2 in. at 3.75 X 100
3 in. at 6.6 X 100
4 in. at 13.7 X 100
5 in. at 24.2 X 100
6 in. at 38.8 X 100

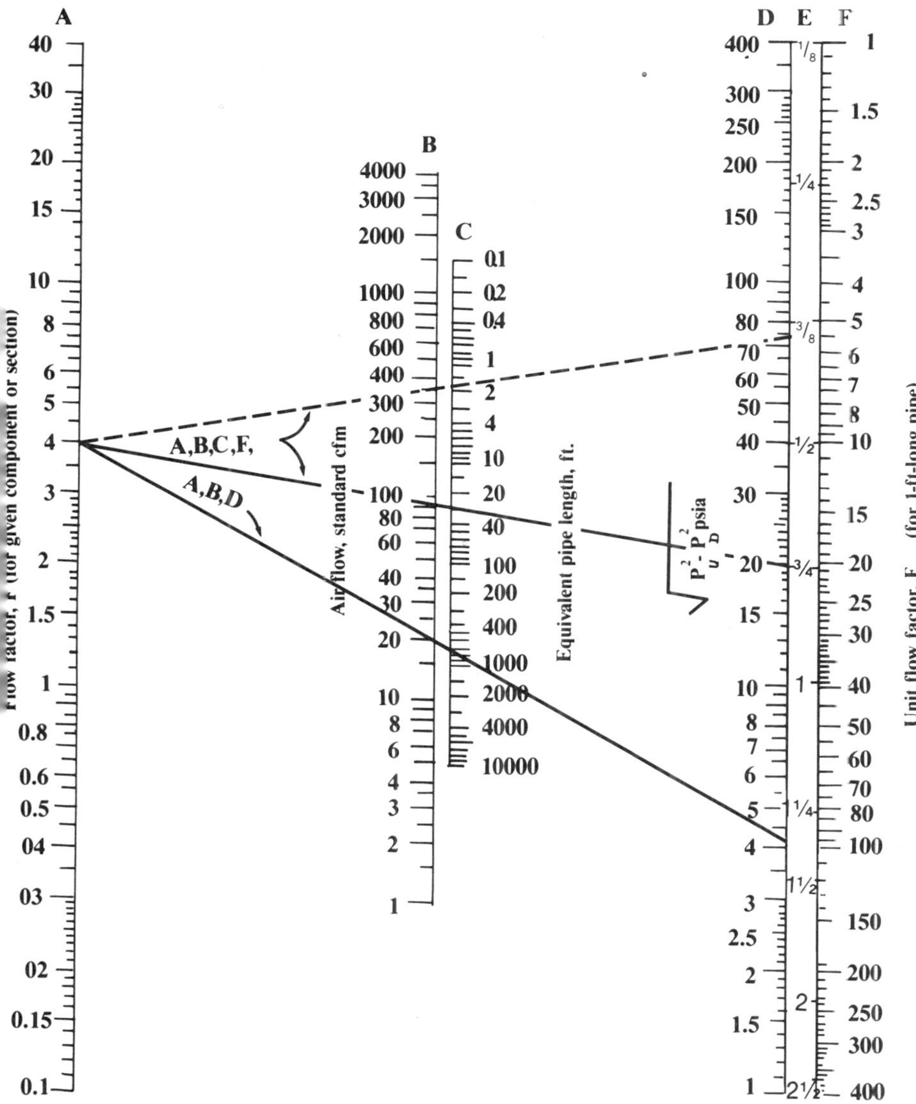

Figure 6.10 Solving for flow-factor pipe length, and pressure loss.

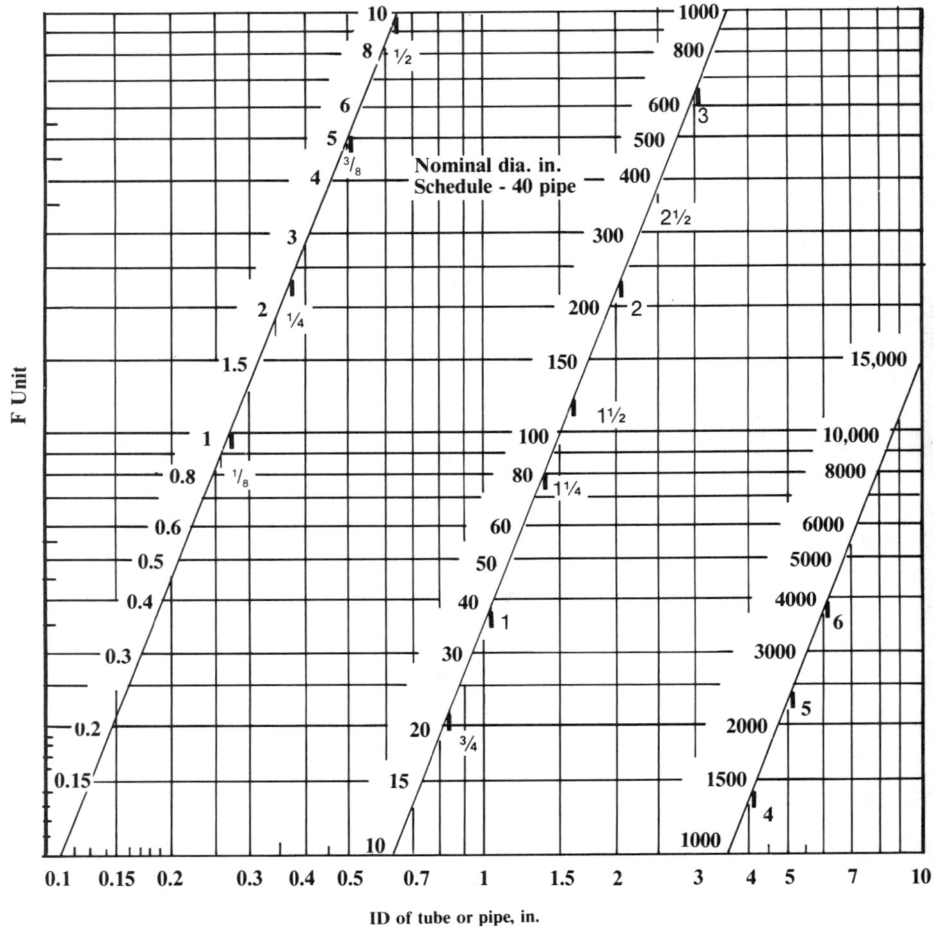

Figure 6.11 Unit flow-factor for tubes and pipes.

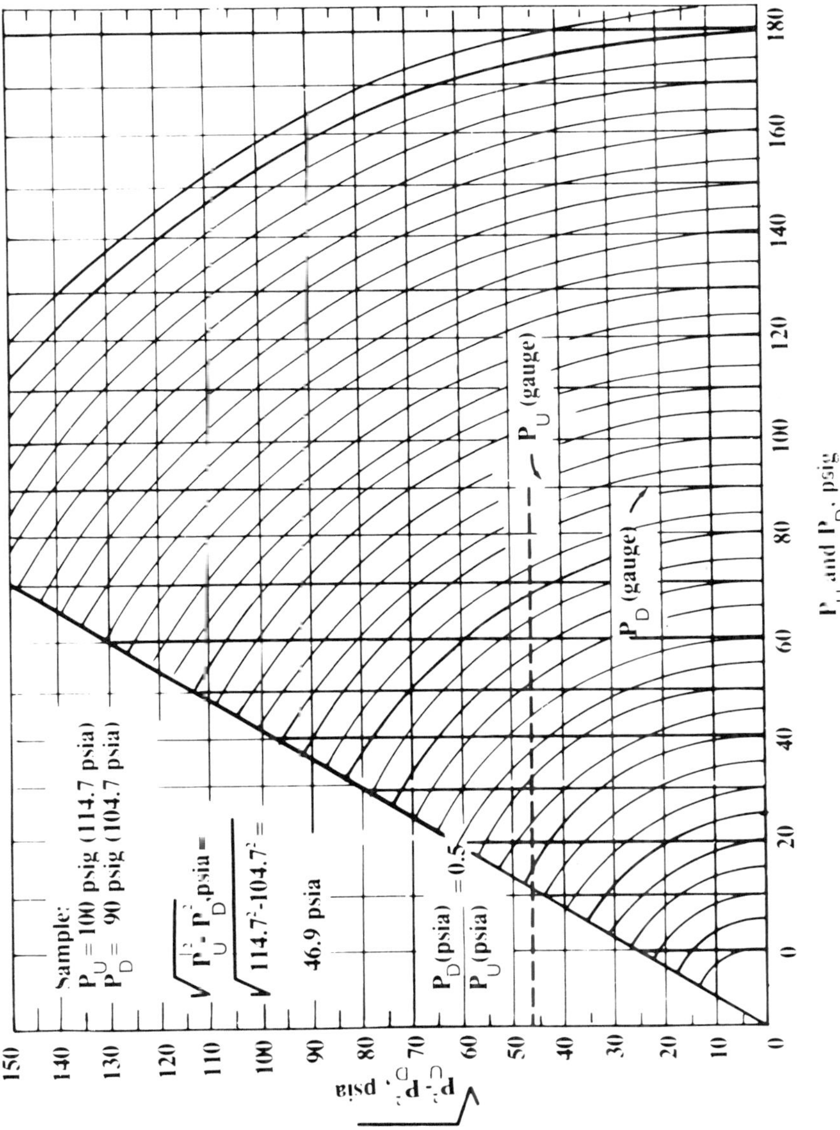

Figure 6.12 Gauge pressures converted to valves for $\sqrt{P_u^2 - P_D^2}$.

At times the pressure loss for a given length of conductor and flow rate is known. If this is the case, the pressure loss can be converted to the flow factor F by using the equation, flow = $F\sqrt{4/3}\sqrt{P_u^2 - P_d^2}$, where P_u and P_d are the upstream and downstream pressure (psia), respectively, and $\sqrt{P_u^2 - P_d^2}$ is obtained graphically in Figure 6.12.

However, in previous chapters we pointed out that most pneumatic valve manufacturers rate their components by something called a capacity coefficient, C_v. We also pointed out that C_v is based on a different air flow equation. But for the case where the ratio of upstream to downstream pressure is equal to 0.5 and the air temperature is 70°F, the C_v equation becomes $C_v = 2.07F$. Now the designer can solve for the flow factor F and then substitute this value directly into the previous nomograph. Steps 2 and 3 will depend greatly on the nature of the design problem. To better understand the design process involved the following example is presented.

6.11. SYSTEM BRANCH ADDITION

Figure 6.13 shows a system schematic with two existing branches. A third branch must be added. This branch will drive an air motor requiring 18 scfm at an inlet pressure of 90 psig. This must be done without disrupting the performance of existing branches.

The solution to this design problem follows the three basic steps outlined above. Tables have been developed that summarize the results of each step. For brevity, calculations and manufacturers' catalog information have been omitted. Likewise, leakage, which may be from 5 to 10% of the flow in a real industrial system, has been neglected.

Step 1 of the design process is to determine the flow resistances and convert them to equivalent lengths of pipe (Table 6.3). Standard catalog data were used for hose, filter, and lubricator. According to the manufacturer's catalog data there will be a 1-psi pressure drop with an inlet pressure of 100 psig at 18 scfm. By using this information and substituting into the flow equation, a flow factor of 1 is calculated. However, the flow factor for the pressure regulator is dependent on the position of its valve element. Therefore, its flow factor is variable. The equivalent lengths

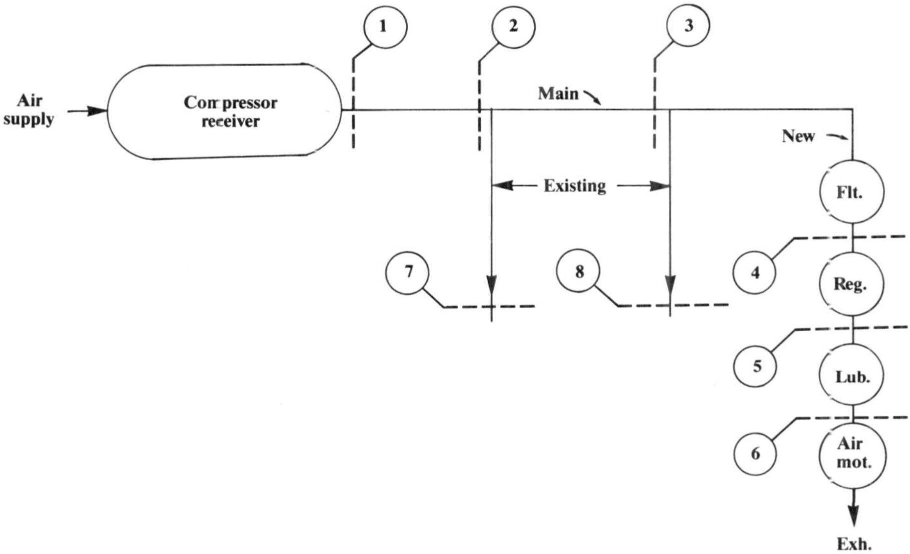

Figure 6.13

of pipe for the existing branches are not discussed because they are known. By using Figure 6.14, equivalent lengths of pipe for fittings and valves can be obtained for the new branches. Pipe diameter conversions are accomplished with the nomograph and scales A, E, C, and F.

The second step is to calculate minimum pressures. Approximate values can be found in Table 6.4. Because separate calculations are made at each tee, there are two values for P_2 and P_3. To satisfy both branch requirements, choose the higher pressure in each case. The flow factor F values were obtained by using the nomograph. Pipe size and length were located on scales E and C. Column A was then read to obtain F. Also, nomograph scales D and B yield values for $\sqrt{P_u^2 - P_d^2}$ and scfm, respectively. Upstream

Table 6.3

Existing section:	Equivalent pipe diameter (in.) (from schematic)	Equivalent length (ft) (from schematic)		Equivalent pipe diameter (in.) (converted)	Equivalent length (ft) (converted)
8 to 3	1/2	54		3/8	16
3 to 2	3/4	35	converted to	3/8	2.2
7 to 2	1/2	27		3/8	7.8
2 to 1	3/4	65		3/8	4.0

New section	Restriction	Flow (scfm)	Pressure (psig) Supply	Pressure (psig) Loss	Flow factor, F	Equivalent diameter (in.)	Equivalent length (ft)
	3/8-in. hose, 10 ft	18	100	1	1	3/8	30.0
	3/8-in. coupling					3/8	0.88

6 to 5	3/8-in. lubricator	18	100	1	1	3/8	30.0
	3/8-in. nipple, 1/3 ft					3/8	0.33
							61.21
	1/2-in. nipple, 1/6 ft					1/2	0.55
	3/8-in. filter	35	80	2	1.56	1/2	40.0
4 to 3	1/2-in. to 3/8. run of tee, reduced					1/2	1.7
	1/2-in. pipe, 30 ft					1/2	30.0
	1/2-in. elbows (2)					1/2	3.4
	3/4-in. to 1/2-in. run of tee, reduced					1/2	0.5
							76.15

(Total equivalent length of section 4 to 3 converted to 3/8-in. pipe: 22.66)

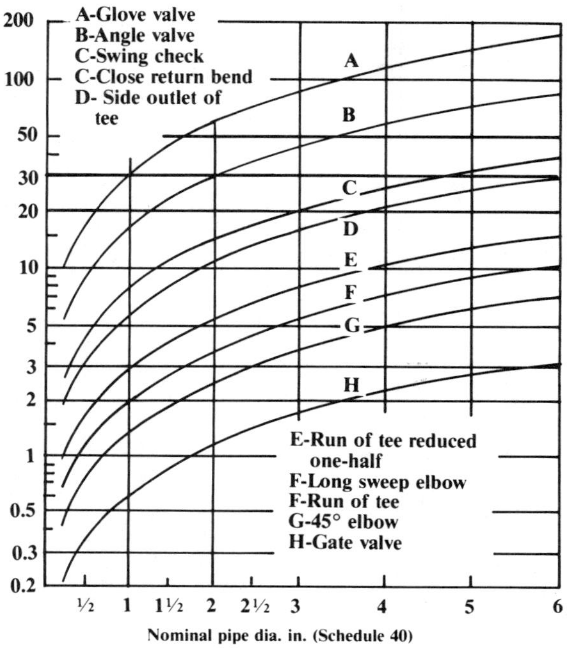

Figure 6.14 Equivalent pipe length for valves and fittings.

P_u and downstream P_d are obtained by using Figure 6.11 and the quantity $\sqrt{P_u^2 - P_d^2}$.

Determination of the actual pressures is step 3 (Table 6.5). Utilizing the nomographs and charts mentioned above, the pressures are recalculated to equal or exceed the required minimum system pressure. These calculations begin at the receiver and proceed to the workstation.

The above method is but one way to obtain a solution to the given design problem. There are other methods that will work on this problem equally well. One such method involves parallel networks. The flow in the main header is divided into three streams, each of which runs uninterrupted from the receiver to a branch. The parallel method is shown in Figure 6.15.

Table 6.4

Section	Equivalent diameter (in.)	Equivalent length (ft)	Flow factor, F	Flow (scfm)	$\sqrt{P_u^2 - P_D^2}$ (psia)	P_D (psig)	P_u (psig)
6 to 5	3/8	61.21	0.69	18	22.5	$P_6 = 90$	$P = 92.5$
5 to 4 Flow regulator—consult flow-regulation chart for pressures						$P_5 = 92.5$	$P_4 = 97$
4 to 3	3/8	22.66	1.14	18	13.8	$P_4 = 97$	$P_3 = 98$
8 to 3	1/2	54	1.36	50	30.75	$P_8 = 105$	$P_3 = 109$
3 to 2	3/4	35	3.55	68	16.7	$P_3 = 109$	$P_2 = 110$
7 to 2	1/2	27	1.93	35	15.7	$P_7 = 110$	$P_2 = 111$
2 to 1	3/4	65	2.64	103	34.5	$P_2 = 111$	$P_1 = 116$

Table 6.5

Section	Equivalent pipe size (in.)	Equivalent length (ft)	Flow factor, F	Flow (scfm)	$\sqrt{P_u^2 - P_d^2}$ (psia)	P_u (psig)	P_d (psig)
1 to 2	3/4	65	2.64	103	34.5	$P_1 = 116$	$P_2 = 111$
2 to 7	1/2	27	1.93	35	15.7	$P_2 = 111$	$P_7 = 110$
2 to 3	3/4	35	3.55	68	16.7	$P_1 = 111$	$P_3 = 110$
3 to 8	1/2	54	1.36	50	30.75	$P_3 = 110$	$P_8 = 106$
3 to 4	3/8	22.66	1.14	18	13.8	$P_3 = 110$	$P_4 = 109.7$
4 to 5	Flow regulator—consult flow-regulation chart for pressures					$P_4 = 109.7$	$P_5 = 92.5$
5 to 6	3/8	61.21	0.69	18	22.5	$P_5 = 92.5$	$P_6 = 90$

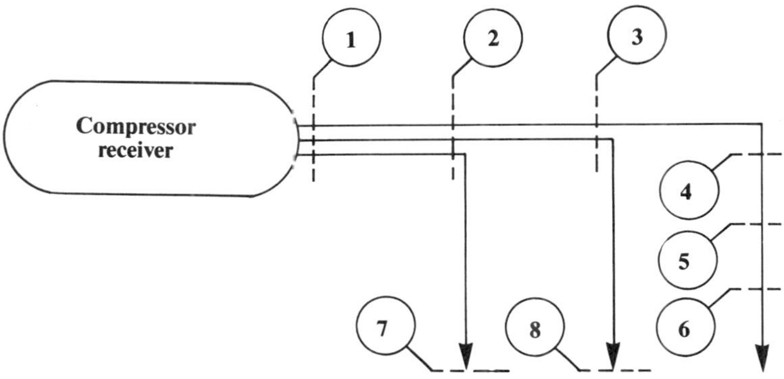

Figure 6.15

This method will work because the flow factors divide in exact proportion to the percentage of total system flow. Also, each branch can be thought of as an independent series system. In this manner, the matching of mixed flows ahead of tees is eliminated.

To illustrate, it is easy to see that the formally single flow path, section 1-2, now has three paths. The lower, middle, and top paths have flows of 35, 50, and 18 scfm, respectively. Also, they have a flow factor F of 0.9, 1.28, and 0.46, respectively. Adding these flow factors together will yield the previous example value of 2.64.

Either of the above methods eliminates the all too often utilized method of trial and error. By using Figure 6.11 to help compute P_d, and by determining the flow factor for each of the parallel sections, the system flow factor can be determined.

7

Installation and Maintenance

7.1. INTRODUCTION

A text on pneumatic valves and circuitry would not be complete without a chapter on the installation, maintenance, and repair of air valves. This chapter is divided into three sections, covering these topics.

Correct sizing and designing of pneumatic valves and circuitry is only the first step in achieving the best circuit design. The best design is one that not only has the correctly sized components properly placed and mounted in a circuit, but also is readily maintained, easy to troubleshoot, and simple to repair. Many impressive circuits fall far short in the latter area. A machine or circuit can be a constant problem if proper installation, maintenance, and repair procedures are not followed. This could lead to excessive maintenance costs, wasted time in pinpointing causes of malfunction, and expensive downtime.

The money and time initially spent on a machine are but a small portion of those spent to keep it "on line." The maintenance cost or time allocation can far exceed that of the initial design project. Therefore, to aid in achieving good design of the pneumatic power system, a guide to valve installation follows.

7.2. INSTALLATION

The proper installation of components in a pneumatic circuit can reduce start-up time, improve reliability, reduce maintenance and operating costs, and optimize efficiency.

7.2.1. System Considerations

One of the first steps is to select hose, pipe, tubing, and tube fittings that will not deteriorate or corrode in the environment of the installation (acid fumes, chemicals, temperatures, etc.).

For proper machine and valve operation the air supply must provide clean pressurized air of adequate volume and pressure. Be certain that the headers and lines from the compressor(s) to the machine are properly sized. Check the available pressure during peak operating periods in the plant. Surge accumulators or storage tanks located close to critical high-flow-demand equipment may be required to maintain pressure. Every fitting, length of tubing, quick disconnect, or other component contributes to pressure drop. The evaluation of the air supply must be made at the actuator or the valve that controls it. The air supply to control the valve must be evaluated as well as the supply to control the working air motors and cylinders.

Figure 7.1

Just as important as the provision of an adequate supply of air is the provision of a readily accessible lock-out exhaust valve to provide a safe means of working on the machine for maintenance or diagnostic purposes.

7.2.2. Manifold/Subbase

If the circuit is such that a need for precise control does not mandate mounting control valves at or close to the actuators, the value and efficiency of subbase and manifold (Fig. 7.1) should be self-evident. In typical applications they allow for the convenient side-by-side mounting of several controls to provide operating efficiency and reduce installation cost by the elimination of many individual tube connections. Simplicity of mounting will reduce time and labor to lower installation expense and prepare for ease of future maintenance.

7.2.3. Miscellaneous

An often overlooked area is that of keeping parts clean. Port plugs should not be removed until valves are ready to be mounted. Do not allow tubing or fittings to become contaminated with shop debris or dirt. To aid in keeping the air supplied to the machine circuit clean, the use of water drains and drop legs is recommended for all pneumatic systems (Fig. 7.2).

Valves and tubing should be rigidly mounted to the machine to control unnecessary vibration. Also, access space must be left for maintenance, manual overrides, and nameplate information. Indicator lights, when provided, should be readily visible.

Although most valves will operate in any position, it is usually good practice to mount spool valves so that the spool is horizontal, preferably along an axis at 90° to the major vibration axis. This will minimize the chance of a valve shifting due to extreme machine vibration and maximize cycle life.

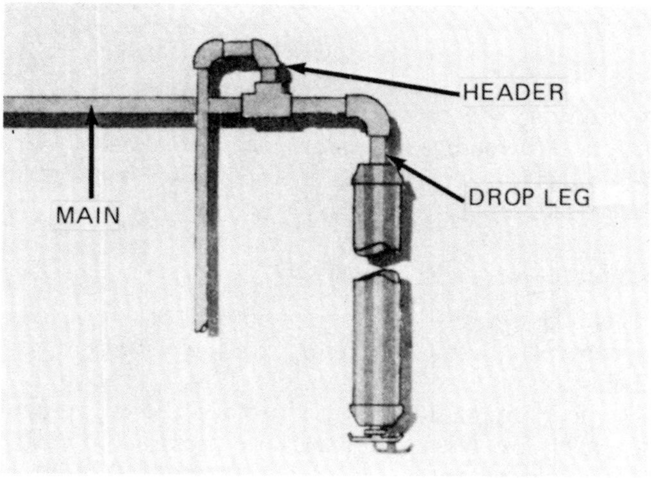

Figure 7.2

Solenoid valves are not the only type found in pneumatic systems. Manual or mechanically actuated valves alone or coupled with air pilot-actuated valves can reduce cost and improve reliability when compared with the electromechanical limit switch, relay, or solenoid-type pneumatic circuit.

7.3. MAINTENANCE

7.3.1. Leaks

Contrary to popular belief, air leaks cost money (Fig. 7.3). Tubing and tube fittings are recommended to minimize the chance of air leaks developing. In smaller valve sizes, especially for 3/8-in. pipe ported (or smaller) valves, the use of plastic conductors with a novel push-in fitting offers assembly convenience and long-term reliability without the use of sealants. However, when pipe is used, a very carefully installed plastic tape or liquid thread sealant is recommended. *Caution*: When using thread sealants, it is imperative that tapes or sealants be applied to male threads only, omit-

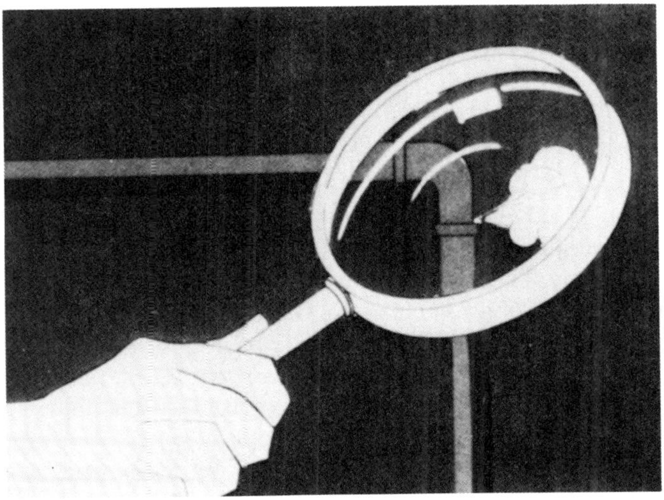

Figure 7.3

ting the first threads, to keep sealing material from being introduced into the system. The new anaerobic thread sealants are preferred for many types of installations.

7.3.2. Solenoids

At the time of installation (and replacement) be sure that the voltage and frequency of the valve coils are correct. Not only during installation, but also during routine maintenance and troubleshooting, it should be determined that solenoid valves are properly connected and that they are *not* running at temperatures in excess of their rating. High resistance in the leads or poor electrical connections may cause solenoid failure. Also, the electrical circuit should be examined to be certain that both solenoids on a double-solenoid valve cannot be simultaneously actuated. For ease of maintenance and flexibility in matching electrical power requirements, "plug-in" solenoids are often used (Fig. 7.4). These solenoids can be rapidly changed to match voltage and frequency and allow rapid testing for possible solenoid failure.

Adverse atmospheres, dirt, water, coolant, and lubricating oils are part of the industrial environment. Valve exhaust ports that face up should be protected with elbows or exhaust mufflers. Also, electrical outlets or access covers should be protected from the possibility of liquids entering the electrical system.

7.3.3. Lubrication

Liquid, generally in the form of oil lubrication, is needed for some circuits and types of controls, but in controlled amounts. Since oil must travel through air lines to reach the valves, excessive run lengths, bends, or angles in tubing should be avoided. When complex piping arrangements cannot be avoided, a recirculating-type lubricator mounted reasonably close to the device needing lubrication is recommended (Fig. 7.5). It may also be desirable to prelubricate the system or components to ensure immediate oil delivery at start-up. To minimize wear in the initial cycles, factory component prelubrication should be considered.

When exhaust air must be free of lubricant or contamination due to the increased cycle life and reliability of a lubricated system, the lubricant-laden exhaust should be piped to a safe point of discharge or properly filtered.

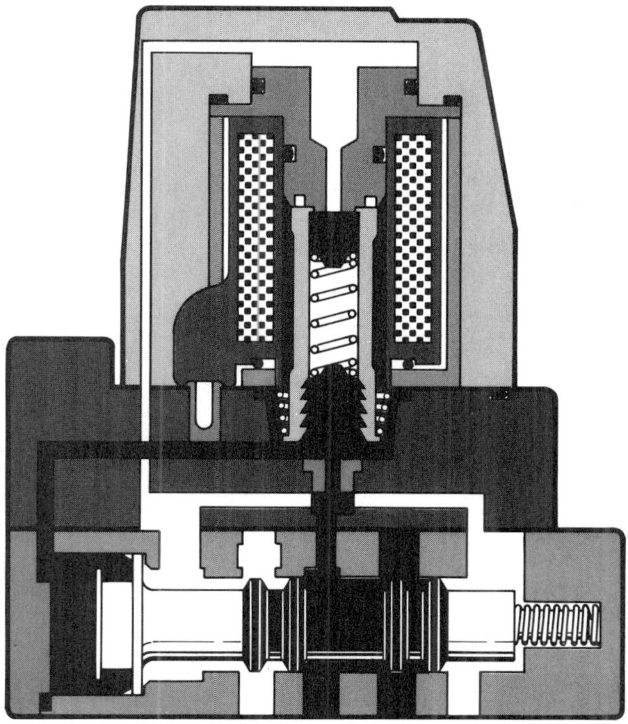

Figure 7.4

7.3.4. Filter-Regulator-Lubricator

To accomplish the above conditions, filters, regulators, and lubricators are recommended for all pneumatic systems (Fig. 7.6). The filter (Fig. 7.7) should be mounted as close to the equipment being protected as possible. This component will trap the water that is condensed and the pipe scale produced in headers and supply lines. Also, there should be no quick disconnects or other points where contamination can enter the system after the filter. Standard filtration is a 40-μm filter element; other options are available, including a 0.03-μm oil removal coalescing type (Fig. 7.8).

As mentioned earlier in this text, air pressure regulation is necessary to eliminate pressure fluctuations in the operation of any pneumatic system (Fig. 7.9). Cost is also important. As a

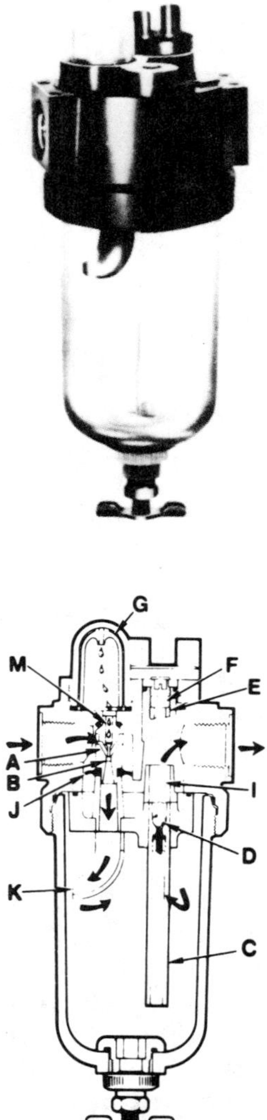

Figure 7.5 A, venturi section; B, throat section; C, pick-up tube; D, check ball; E, metering block assembly; F, metering screw; G, sight dome; I, mist particle opening; J, restrictor disc; K, curved baffle plate; M, nozzle tube.

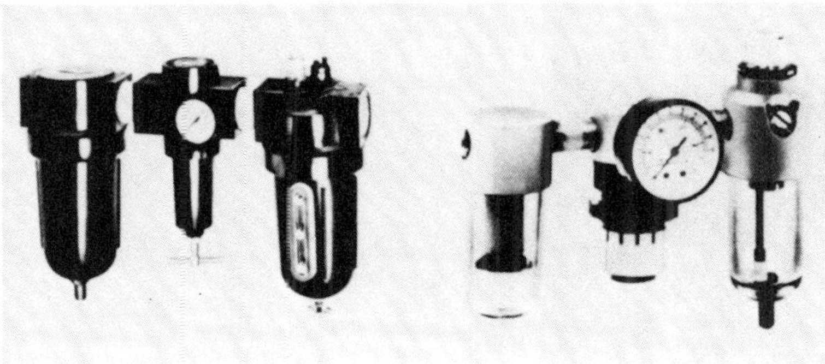

Figure 7.6

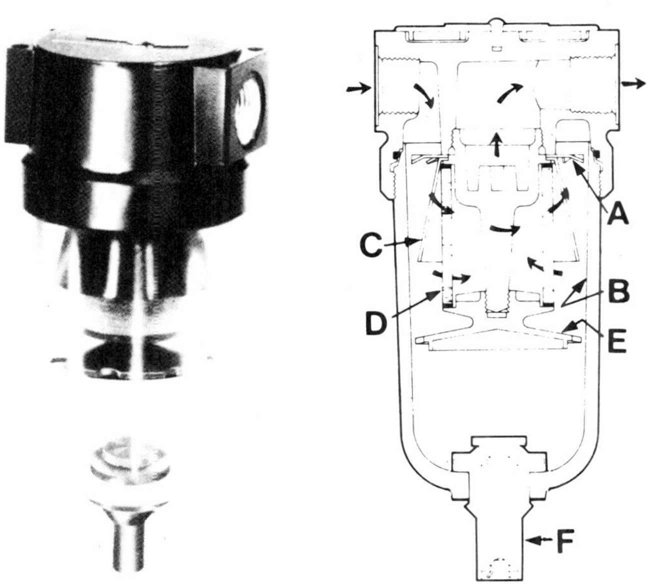

Figure 7.7 A, deflector plate; B, interior bowl wall; C, shroud; D, filter element; E, baffle; F, manual drain.

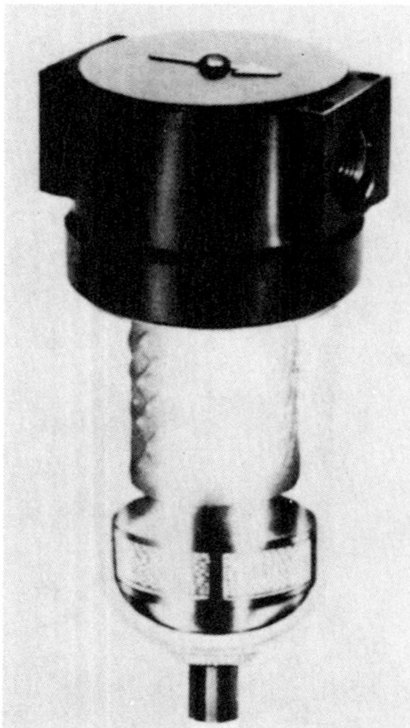

Figure 7.8

rule, air at 100 psi costs twice as much per cubic foot as air at 50 psi. The pneumatic system should be operated at the minimum pressure required to get the work done. Dramatic reductions in operating expense have been achieved by using dual pressures (high pressures where high forces or accelerations are needed and low pressures where they are sufficient for low-power-demand portion of the machine cycle).

The in-line mist-type lubricator (Fig. 7.10) is probably the most inexpensive and accurate of the lubrication devices available. The lubricator should be mounted as close to the point of work as possible. The choice of lubricant is determined by its compatibility with system components and seals. In the case of components requiring minimal air flows, such as small cylinders, a lubricator may have to be mounted close to the actuator, that is, between the valve and cylinder.

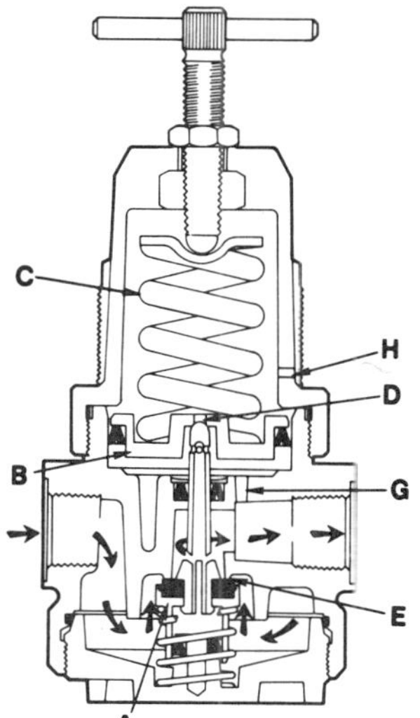

Figure 7.9 A, poppet assembly; B, piston; C, control spring; D, piston vent hole; E, seat; G, pilot orifice; H, bonnet vent hole.

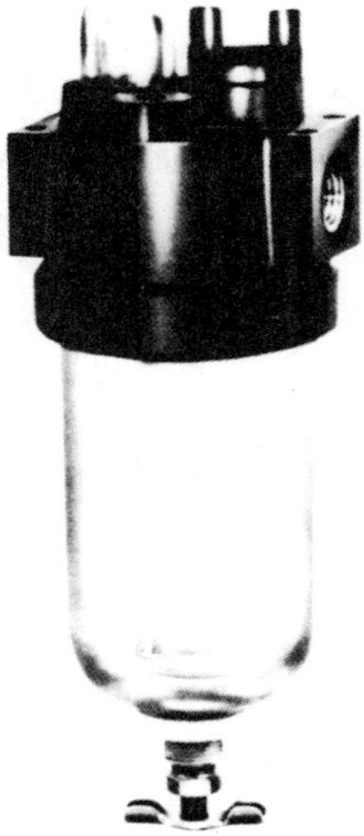

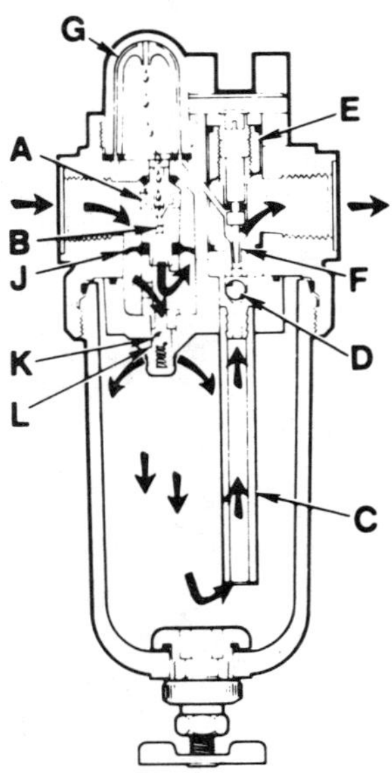

Figure 7.10 A, venturi section; B, throat section; C, pick-up tube; D, check ball; E, metering block assembly; F, metering screw; G, sight dome; J, restrictor; K, check ball, L, check ball seat.

7.3.5. Accessories

There are other pneumatic system accessories, and to gain the optimum cycle rate and maximum control of the pneumatic system, each accessory msut be properly installed. Flow control and quick exhaust valves should be mounted at the cylinder ports for optimum control.

Accessories such as mufflers and silencers (Fig. 7.11) should be protected or mounted in a manner to prevent contamination and pressure buildup. Either or both will hamper system performance. Internal contamination as well as external contamination will cause the system to malfunction by preventing the valve from shifting. These accessories should be serviced regularly and replaced periodically.

7.4. REPAIR

In time it may become necessary to remove and repair a malfunctioning valve. When this situation occurs, consult the machine manufacturer's maintenance instructions. The first step in the removal and repair should be to shut off the air supply to the valve or leg of the circuit. This can be done by closing a lock-out shutoff valve (Fig. 7.12). This valve closes off the air supply and opens the machine's lines to exhaust. It has provision for a padlock to secure the valve in the off and exhaust position. It is an unsafe practice to work on components and/or circuits that are pressurized.

Standard procedure in many companies is to isolate the malfunctioning component and replace it with a new or repaired one. This puts the machine back into production quickly. The malfunctioning component is then disassembled and repaired in accordance with the manufacturer's recommendations, if available.

When repairing a malfunctioning component, all damaged, failed, worn, sticking, or broken parts must be examined and the reason for the failure isolated and removed. Referring to Chapter 8, on troubleshooting, will aid in this area.

The valve manufacturer's repair and replacement procedures should always be followed when servicing a component. There may be construction or assembly features that must be considered and provided for to effect a safe repair.

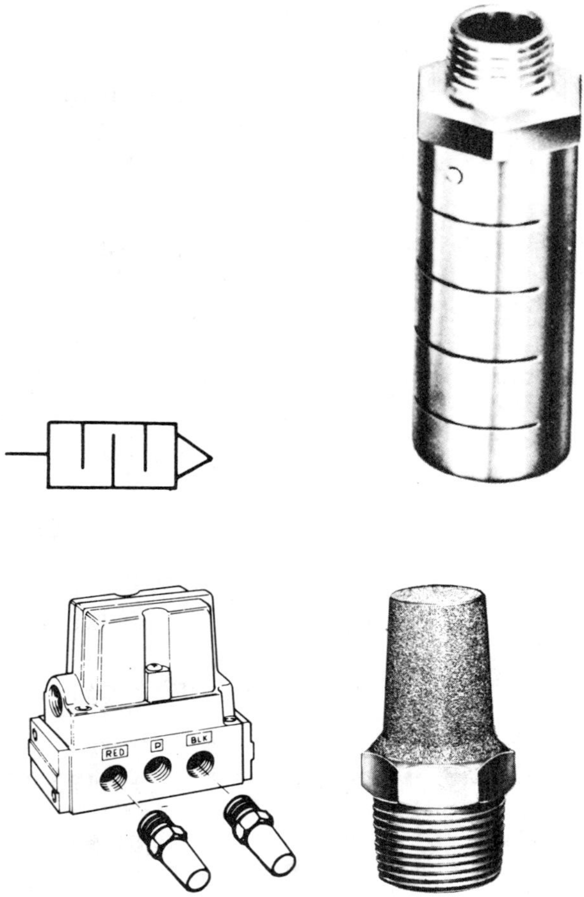

Figure 7.11 Top left, silencer (muffler) graph symbol; top right, silencer; bottom, exhaust port mufflers.

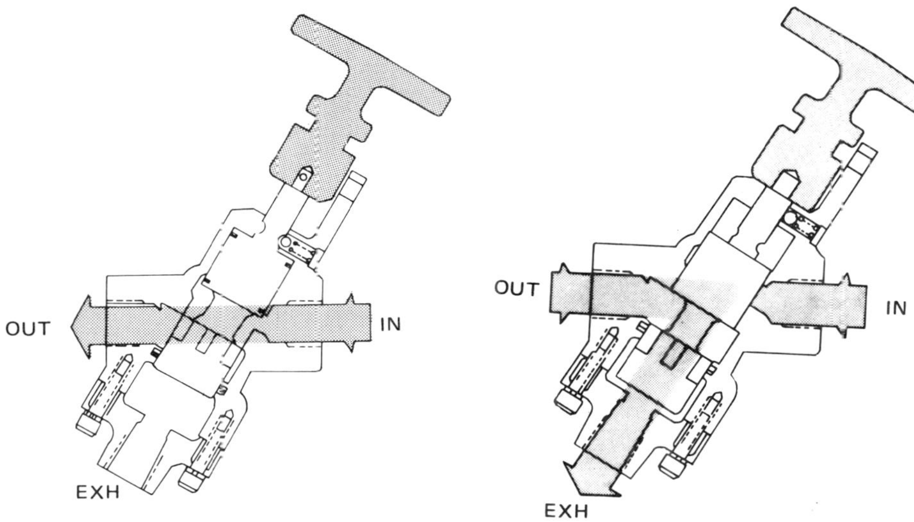

Figure 7.12 (Courtesy of Ross Operating Valve Company, Detroit, Michigan.)

Also, the repaired component should be properly tested, again according to the manufacturer's recommendations, before it is put back in stock or used again. It is this pretest that will determine whether the valve has been repaired properly. If the valve is not properly tested, but just put back into stock after repair, its true operating condition will not be known. In fact, the reason for the malfunction may not have been corrected. This can lead to excessive downtime when a malfunctioning valve is replaced with one believed to have been repaired. It also complicates troubleshooting, something that is not needed in standard or unusual pneumatic circuits.

8

Troubleshooting and Repair

8.1. INTRODUCTION

Being able to design a pneumatic system, select the correct components, and deliver a complete system is only a small portion of the design responsibility. The useful life of a machine is directly related to its maintainability. As an example, most air valves are reliable, rugged, and dependable pneumatic devices. They find use where high-speed, repetitive operations are needed, such as on automatic production equipment. The service life and performance of the air valve are based on proper application and preventive maintenance. However, malfunctions and failures do occur. When a pneumatic circuit malfunctions, the thoughtful troubleshooter first seeks to study the circuit diagram schematic normally provided by the machine designer or builder. If the problem appears to be caused by a specific component, the maintenance instructions for that component must be referred to and followed in putting the unit back in working order.

The following is intended as a supplementary guide for correcting frequent problems that are encountered with air valves or as a general guide if the valve manufacturer's data are not available. In many cases, observation of a symptom or a measurement with standard shop maintenance test equipment will supply the information leading to the solution of a problem.

Although we have focused on the analysis and correction of valve component malfunctions, we believe that the process of pinpointing the cause of malfunctioning pneumatic circuits can be greatly simplified. One step in this simplification process is through the installation of diagnostic access devices (diagnostic fittings). These should be considered for circuits that are adjusted for specific tasks or those which have a tendency to malfunction.

In a pneumatic circuit, the addition of quick coupler female valves (those designed for pneumatic system use) at key points of the circuit should be seriously considered. This will allow rapid checking of pressures, at different portions of the circuit, in much the same manner that voltages can be checked in electrical control circuits.

8.2. SOLENOID FAILURES IN GENERAL

A solenoid is an electromechanical device that converts electrical power or energy into mechanical force and motion. Solenoids used to operate pneumatic valves may fail for a variety of reasons. The following is an explanation of the more common reasons for solenoid failure.

8.2.1. AC Solenoid Failure

A solenoid designed to work on alternating current (AC) has a relatively low electromagnetic resistance in its initial or normal "open" position. Because of that, it allows a large flow of inrush current. As the solenoid's armature moves, it shortens the magnetic path. This increases the electromagnetic resistance of the solenoid and since the applied voltage is essentially constant, the current decreases as the magnetic path is shortened. At the end of the solenoid armature's stroke the current is at a minimum—dictated by the resistance of the coil [ohmic (DC) and inductive]. Therefore, the initial (inrush) current is several times as large as the final (sealed) current.

Typically the solenoid armature is in contact with a spring return (spring-biased) valve member. As the alternating current goes from one peak, falls to zero, and rises again, the solenoid plunger would alternately be forced away from its seat by the bias spring when the current passed through zero and then pulled back on its seat at peak current points. A loud AC "hum" would result. This situation is controlled by the use of a "shading coil," which helps keep this noise under control.

To put the above in perspective, remember that an AC solenoid-operated valve usually completes its solenoid-driven stroke in 20 to 50 ms. From this it can be seen that delays in shifting as well as high rates of cycling (such as AC hum) cause unusual inrush power conditions to exist and heat up the coil.

The "power" or pull-in force of a solenoid is proportional to the square of the applied voltage. The relative pull-in force for a given solenoid on a valve operating at a voltage other than its design value may be estimated from:

$$\text{Force (actual)} = \frac{\text{volts}^2\text{ (actual)}}{\text{volts (design)}} \times \text{force (design)}$$

When the line voltage of a solenoid designed to operate at 115 V drops to about 100 V at the solenoid the available solenoid power drops to 76% of the designed force, and not enough magnetic force may be developed to allow the armature or plunger of the solenoid to move. If the solenoid plunger cannot move, or can only move part of the way, this will result in a continuous high inrush current draw for an abnormally long period. This causes the solenoid to operate at temperatures higher than it was designed for and the solenoid fails or burns out.

Voltages should be checked quickly at the coil with the solenoid energized and held in its open or de-energized position if possible. This checks the solenoid circuit at its maximum current demand condition and shows whether there is a high resistance in this circuit. Note that any delays in taking the inrush measurements can create considerable heating of the solenoid and possible solenoid failure. The possible causes of low voltage under load include broken strands on connecting wires, high-resistance connections, excessive loading on the electrical circuit, and low voltage at the control transformer that powers the solenoid.

If the solenoid armature or plunger is not allowed to move or seat because the valve spool will not shift, high inrush current will continue until the solenoid burns out. This type of failure is typical in directly operated solenoid valves. Directly operated valves generally have larger solenoids than pilot-operated valves. The directly operated air valves may have electrical power requirements as high as 35 W or more, usually well over 10 W.

A common type of directly operated solenoid valve is the lapped spool design. In this configuration resilient seals between moving parts are eliminated; they are not required because sealing is accomplished by the very close fit between the spool and valve body. Because of this tight fit the spool may varnish in place, or airborne contaminants may become wedged in and prevent the spool from shifting. The valve should be cleaned, lubricated in accordance with the manufacturer's maintenance instructions, and reassembled. If the varnishing problem persists it may be necessary to place an oil-removing type of in-line air filter ahead of the valve. This often clears up the problem as long as the filter is serviced properly. The filter will require periodic checking and element replacement.

When resilient seals are used in the valve, as in the packed spool and packed bore, the valve spool may not shift because of swollen or sticking seals. When malfunctioning components are encountered or isolated the manufacturer's maintenance instructions should be consulted first. If examination shows that the seals are swollen or spongy, the problem could be due to synthetic, low-aniline, or high-flash-point or even fire-resistant phosphate ester lubricants being used in the compressor, or the wrong lubricant being used in an in-line air lubricator. The seals and/or the spool should be replaced and placed into the cleaned and lubricated bore. If the seals are sensitive to the type of lubricant found in the system, the valve manufacturer's recommendations should be followed.

If both AC solenoids of a double-solenoid, directly operated valve are energized at the same time, one solenoid will seat before the other. This will prevent the second solenoid from seating, subjecting it to a high inrush current and ultimately burning it out. To check for this problem, small indicator lamps should be temporarily wired across the coil of each solenoid. If both lamps are ever lit at the same time there is a problem in the electrical control circuit which must be corrected. The machine manufacturer should be consulted for a recommendation before modifying the circuit. If the electrical control circuit cannot be corrected or made foolproof, the use of a pilot-operated, solenoid-controlled valve should be considered. In this type of valve the solenoids do not exert their force against each other. If a change to a pilot-operated valve is made the circuit may work, but be unsafe. A change should not be made without a detailed study of the probable results.

Higher transient voltages will cause an increased or extra solenoid force. However, the overvoltage is typically of short duration. The mechanical life of the solenoid may be affected by this overvoltage, but the added force generally does not cause a solenoid failure. If the overvoltage is very high and of long duration, it will cause breakdown of the coil insulation, especially where individual turns of the solenoid winding may cross another turn at a different voltage potential, resulting in a turn or two of the winding becoming short-circuited. This overvoltage problem is most common when a solenoid is connected to a line of greater voltage than it is designed for, or where high-voltage transients occur.

A major cause of this problem occurs when the solenoid is connected without isolation to an electrical circuit that also delivers power to induction motors or other inductive load-type equipment. Switching in and out of this type of equipment can cause very high overvoltage peaks in the circuit.

Control circuits should be isolated from the main power circuit and protected by the installation of electrical filter networks. Industry recommendations concerning power levels and grounding should always be followed.

Often loud AC solenoid humming is an indication that something is wrong with the solenoid or its application. If the solenoid noise level is high when the solenoid is energized, failure of the armature or plunger to seat should be suspected. Most directly solenoid-actuated valves are provided with manual overrides. If the noise decreases when the override is operated, incomplete solenoid motion is confirmed. Solenoid voltage should be checked to see that all moving parts are free to move.

A broken part inside the solenoid may cause an extremely loud AC hum. This calls for replacement of the solenoid.

The solenoid designer incorporates a shading coil to minimize the humming. This coil is a copper ring mounted in the solenoid frame to provide an induced current that lags the applied current. This induced current flowing through the low-resistance, single-turn shading coil is sufficient to prevent the armature from moving when the current passes through zero as it proceeds from one maximum value to the next. The shading coil prevents the solenoid load (most often a spring) to push the armature or plunger away from the seat. A broken or dislodged shading coil will allow the spring to move the plunger away from its seat, causing the solenoid to hum loudly because of the repeated (120 Hz) seating impact. If this is the problem the solenoid should be replaced.

In applications where low noise is required, mounting the air valve so that the solenoid operates vertically should be considered. If maximum quiet is required, DC solenoids should be considered.

If the solenoid is operated in an environment where the ambient temperature exceeds the maximum rated solenoid temperature, solenoid failure may occur. The insulation used on solenoid wire is rated by class; the classes are A, B, F, and H. Each class has it own characteristics. Class A is the lowest rated with a maximum operating temperature of 221°F (105°C). The highest rated is class H insulation, with a maximum operating temperature of 356°F (180°C). If a coil is operating over its maximum temperature rating, the winding will overheat and eventually burn out. If high-temperature insulation is not available, or if the valve does not have such insulation, the valve must be placed in a cooler location. Another alternative is to replace the valve with a remotely piloted valve. Extreme low-temperature operation in an environment with an ambient temperature lower than rated may also cause solenoid failure. The dissimilar metals of a valve may have sufficient differences in their coefficients of thermal expansion to cause them to bind. Lubricant viscosities will increase, retarding or stopping solenoid motion. This problem can often be resolved by using a lubricant suitable for low temperatures (e.g., silicone lubricants), moving the valve to a warmer place, or providing a heated enclosure. If this problem continues the valve should be replaced by one that is designed for low-temperature operation.

Extended exposure to contamination may result in solenoid failure. This happens if the solenoid is left uncovered. Metal chips collect in its magnetic field and can limit or stop armature movement. The result may be a constant high inrush current and ultimate solenoid burnout. This problem is rare because the Joint Industrial Council Standards require that air valves become inoperable when the solenoid cover is not in place. However, even covered solenoids may fail due to contamination. If solenoid valves must operate in adverse environments containing dust or chips, or where they may be splashed with water coolants, they should have sealed electrical connections and the solenoid cover should be kept tightly in place.

To check the condition or continuity of a solenoid, a calibrated ohmmeter is required. First the coil resistance of a similar solenoid coil should be determined. Then one of the valve solenoid leads should be disconnected and a meter reading taken across both solenoid leads. If the good and the problem coils are the same temperature and the meter shows a resistance which differs from that of a known good coil, the solenoid circuit may be complete but a problem may still exist. Perhaps an incorrect coil has been used. The coil part number, usually marked on the coil itself, should be checked. If the resistance reading is lower than that for a known good solenoid, shorted turns may be the cause. If the meter reading is zero the solenoid is shorted. Then, with the ohmmeter set at the highest resistance scale, readings should be taken between ground and each solenoid lead. If the reading is zero or close to it, the solenoid is shorted to ground. High temperature or overvoltage could be the cause. A reading of infinity indicates a broken wire or open circuit. External mechanical shock or vibration could be the cause.

8.2.2. DC Solenoid Failure

Direct-current solenoids do not experience high inrush current (compared to sealed current) as do AC solenoids. Therefore, DC solenoids should not burn out due to low voltage or failure of the armatures to seat. As with AC solenoid valves, the DC solenoid valves can fail to shift. Failure of a DC solenoid valve to shift may be caused by an electrical problem. The most common problems of DC solenoids are similar to those that cause failure in AC solenoids. Lead wires may have broken strands, creating high resistance and thus lower voltage at the solenoid. Coils may have been manufactured with a latent defect—for example, poor electrical connection to the coil wires; shorted turns; wire almost broken; solenoid wire wound too tightly, causing wire to be stretched and thus have too high a resistance; wire wound too loosely, causing internal motion and shorting of turns.

But failure in a large DC solenoid, which draws in excess of 15 to 20 W, probably is a result of contactor problems because of the heavy inductive arc across the contacts each time the electrical circuit is interrupted. Such a failure often appears intermittently and thus is very difficult to pinpoint. If large-current-draw DC solenoids are used, the contactors should be clean. Normally, they will be protected electrically against excessive arcing. Two common methods of protecting open contactors are shown in Figures 8.1 and 8.2.

Corrosion and temperature can also affect DC solenoid operation. The general guides for AC solenoids can be applied in these cases. High transient voltage problems can be handled in the same way as those of AC solenoids.

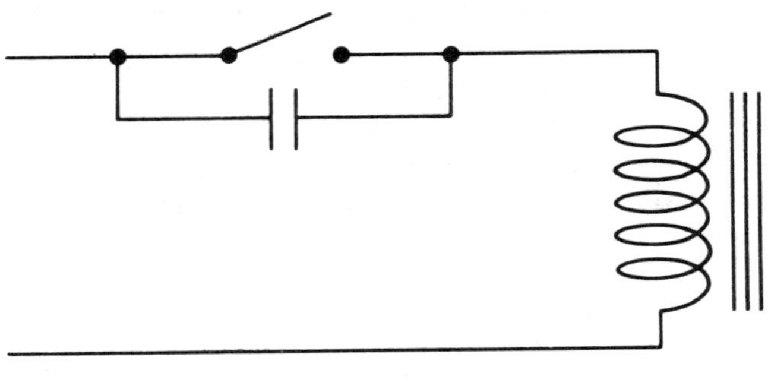

Figure 8.1 Capacitor used to suppress spark.

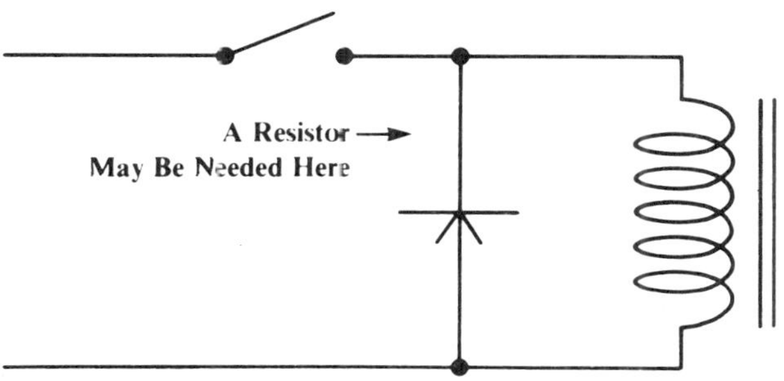

Figure 8.2 Rectifier used to suppress spark.

8.3. SOLENOID-CONTROLLED, PILOT-OPERATED VALVE FAILURES

Because of the limited shifting force that is developed by a reasonably sized solenoid, only smaller air valves are of the direct-acting solenoid type. In large air valves, flow and pressure are directed through a smaller air valve to poppets on either side of the larger valve's spool to shift it. These solenoid-controlled, pilot-operated valves require considerably less electrical power or energy.

8.3.1. Supply Pressure

For the pilot-operated valve to operate, a minimum pressure must be available to overcome such forces as seal friction, a spring, or detents. Typically, the minimum pilot pressure could be as low as 15 psig (1 bar) or as high as 30 psig (2 bar) or more. Pilot pressure checks should be made at the valve inlet port. If the pressure is found to be below the minimum required and it cannot be raised to the minimum or higher, then a valve that will operate reliably at the available pressure should be installed or a directly operated valve or a remote pilot option for the valve should be considered.

8.3.2. Partial Valve Shifting

Internally piloted valves may partially shift and then stop or stall, allowing air to blow steadily through the exhaust port. The escap-

TROUBLESHOOTING CHECK LIST FOR PNEUMATIC PILOT-OPERATED DIRECTIONAL-CONTROL VALVES

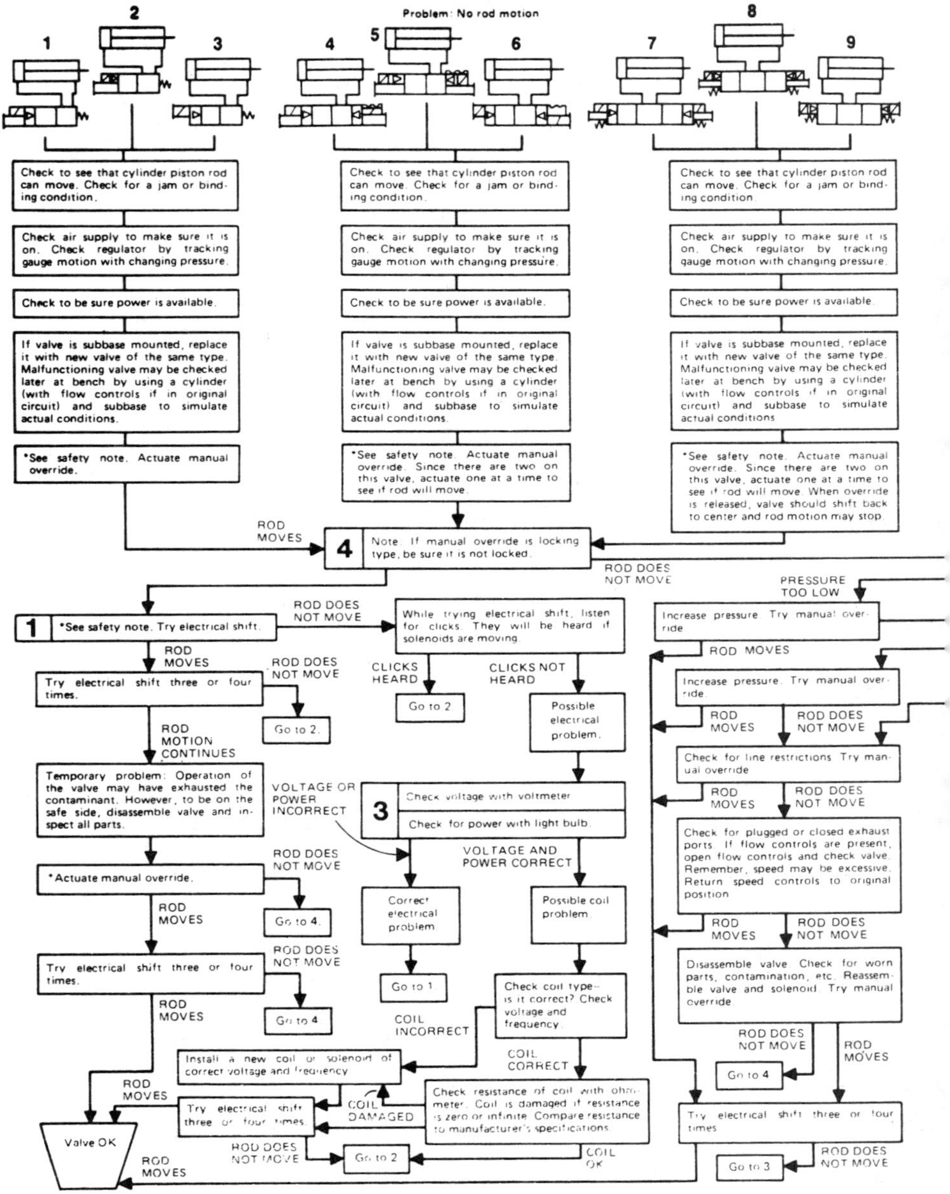

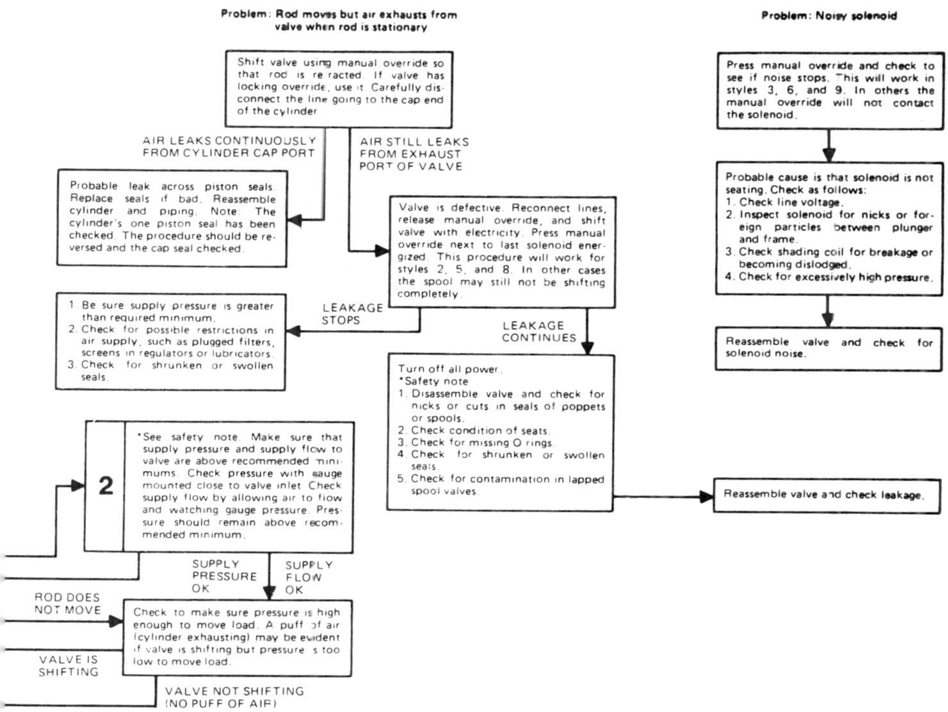

Figure 8.3 Troubleshooting checklist for pneumatic pilot-operated directional control valves. Troubleshooting is perhaps the most challenging aspect of fluid-power work; it is as much an art as a science. As in any art, there are no substitutes for common sense and experience. This flow chart will, however, give maintenance personnel a good start—an indication of what is likely to be wrong and how to repair it. The user should begin by identifying the appropriate valve design at the top of the chart. *Safety note. Before tackling any troubleshooting problem, be sure that the actions taken will not cause damage or injury to machine or personnel. Visually inspect the machine, check the services to the machine, and determine at what point in the cycle the machine has stopped or malfunctioned. When working with electric, air, or hydraulics, be sure that power to the machine has been locked out. (Copyright 1980 by Parker Hannifin.)

ing air is an indication that the pressure and exhaust are connected while the valve element is in motion. This is a characteristic of some pilot-operated, poppet-style valves. The escaping air is a sign that the inlet port pressure has fallen below the valve's minimum operating or pilot pressure. During the poppet's short shift, a very large flow path is opened from supply to exhaust. In some cases this "crossover" flow may deprive the valve of the needed pilot pressure to complete the shift. If the valve design permits, a small accumulator in the pilot air circuit will provide a certain supply of air to allow the pilot circuit to function reliably [U.S. Patent 3,038,500 (expired) shows this]. Providing external exhaust for the pilot air pressure, increasing the supply pressure, and providing a local air tank to maintain the pressure level at the valve during periods of high flow are possible solutions.

In addition, restricted supply lines will cause low inlet port pressure. Supply lines should be checked for such common restrictions as clogged filters, restrictive lubricators, undersized, kinked, or collapsed hoses, and defective fittings or line joints. Conductors and fittings should be sized with internal cross-sectional areas as large as the inside diameter of the pipe for which the inlet port is tapped. If the valve is provided with an internal pilot exhaust, it may be desirable to change it to an external exhaust if possible.

8.3.3. Restricted Exhaust Port

Depending on the style of valve used, restricting the exhaust port with such things as flow control may create a back pressure in the valve, preventing a complete shift of the valve's element. Manufacturers' recommendations should be followed when exhaust port flow controls are being considered.

Restricted valve exhaust can also be caused by flow controls or mufflers becoming plugged with ice crystals. Wet air or air with a sufficient percentage of water vapor tends to freeze as the rapidly exhausting air cools the moisture below its freezing point. This condition is most common when the ambient temperatures are low, but not freezing. If outside temperatures are low, below freezing, then the moisture in the air taken into the compressor is very low and freezing of exhaust is rare. If the valve occasionally malfunctions and circuit reaction is slow, the exhaust ports of the valve and/or muffler should be checked for signs of icing. A solution to this problem would be to dry the incoming air, lower its dew point, use a different muffler, or heat the areas that tend to freeze.

Figure 8.3 is a flowchart to aid in troubleshooting pneumatic pilot-operated air valves.

8.4. GENERAL PNEUMATIC TROUBLESHOOTING TIPS

Troubleshooting air valves is only one aspect of pneumatic system maintenance. In fact, troubleshooting is perhaps the most challenging aspect of such maintenance. A machine sometimes seems to mask the real problem, or deceives the troubleshooter with a variety of symptoms. However, one of the aspects of dealing with a machine is that there is always a reason for any malfunction. The job of the troubleshooter is to find it. Usually it is a very simple fault. Fortunately, a machine in trouble will provide the troubleshooter with all types of clues to the reason for its behavior. All the troubleshooter must do is interpret the clues. Most troubleshooting occurs in either start-up or breakdown situations. Each case will privide its own set of clues and require slightly different techniques on the part of the troubleshooter.

Regardless of which situation exists, certain general rules always apply. First, obtain and study the circuit drawing and maintenance instructions. Then:

1. *Take nothing for granted.* Make certain that a safe condition exists. Do not take the word of someone else regarding a critical item, such as determining whether there is any "juice" on the machine. They may not know what they are talking about. Someone could be electrocuted, so check this out for yourself.
2. *Keep both hands in their pockets.* Resist the temptation to dive in. Resist any pressure or comments such as, "Don't just stand there. Do something!" until you are absolutely certain of what the results will be.
3. *Keep calm.* The machine has the advantage in this respect. It does not have emotions. It will chop a hand off without any show of emotion.

4. *Think before acting.* This is the most important rule, and the one most often overlooked.

8.4.1. Troubleshooting a Start-Up

More machine malfunctions occur during start-up than at any other time. To begin troubleshooting, a current accurate drawing of the circuit and a clear head are needed. Depending on one's familiarity with the machine, the following visual and mental steps should be taken as required.

1. *Visually inspect the machine* to become familiar with its mechanisms and general layout. Ask as many pertinent questions as possible from anyone present who is in a position to add information.
2. *Check all services to the machine.* Is there electricity of the proper voltage, current rating, and frequency on the machine? Are there any high-pressure steam lines, gas lines, or any power devices that require special attention? If in any doubt, check!
3. *Check the control drawing* with the machine. (*Note*: In this chapter such terms as control drawing and/or control circuit mean that portion of the system which is performing the logic functions.) Where are the push buttons? Take special note of any emergency controls shown on the drawing. Know where they are. At this point do not trust them to work.
4. *Check the power elements with the circuit diagram.* Are these elements in the stated "initial conditions" required by the drawing?
5. *Check the control system status.* Are the limit valves made or released as required by the initial conditions shown on the drawing? Were any damaged in transit? Does this system have a control "reset" feature?

After this checklist has been completed, there will probably be several discrepancies. The question is, how can they be corrected?

So far nothing should have moved on the machine. Now something must. Thousands of dollars in tooling and machinery may ride on the next decisions. Think before acting. Take every precaution. Take one simple step at a time while analyzing the result of each step in isolating the malfunctioning component.

Work carefully and methodically, one simple step at a time, as follows:

1. *If possible, isolate the power air* to the leg of the circuit that is malfunctioning. [*Note*: In this chapter terms such as power air and/or power circuit mean that portion of the system which is not performing the logic function(s), but instead is being used to perform work (power).] This could be done by using lock-out valves. A lock-out valve is a three-way device that is used to block off the air supply to a portion of a circuit and open that part of the circuit to exhaust. It is so constructed that its movable element can be externally locked in the nonpassing/exhaust position. This will prevent someone from accidentally turning on line pressure to a circuit leg while it is being serviced. Set pressure as low as possible on the circuit regulator to prevent any damage should slides and so forth be out of position. Now manually cycle the manual overrides to determine whether the power valve piping is correct and flow paths are unobstructed. It may be useful to remember that all passages in fittings, hose, pipe, and other components may *not* be open. They may not have been manufactured properly or they may have been installed improperly so that an air passage is restricted or blocked.

2. *Reset all power elements to initial conditions as shown on the circuit diagram.* Check to see that the associated limit valves are properly actuated or released as required. Resolve all discrepancies as they occur with the air off. Never work on a system or component without first setting and locking the system lock-out valve in the nonpassing condition. Also, check the freedom of movement of all manual controls with the air off. Turn the pressure up to the required level. Listen for leaks. Next, adjust flow controls for the lowest practical speed by cycling with the manual overrides. Finally, reset the power elements to their initial conditions.

3. *Set all push-button selectors for the simplest possible cycle or operation mode.* Normally all selectors would be set to

conform to the "initial conditions" statement on the drawing. This would put the machine in a "manual" mode so that the available manual push buttons would be armed. If no manual push buttons are available, choose the simplest cycle possible.

4. *Turn control circuit on.* If anything moves, turn the system off immediately and find out why. If nothing moves, listen for leaks. One distinct advantage of air over electricity is that a leak is audible and not likely to knock you across the room. (Few people are pneumaticuted.) Leaks will noisily direct the troubleshooter to a poor connection. Air hissing out of an exhaust port will immediately identify a malfunctioning component.

5. *Operate remote manual push-button controls.* If all power valves respond as required, this proves that the hookup is correct, at least as far as the manual side of the circuit is concerned. It is also comforting to know that the circuit can later be switched to the manual mode to get out of a sequence problem. Make sure all power elements are reset to initial conditions before proceeding.

6. *Reset control circuit to initial conditions.* This circuit should be designed with a master relay reset to facilitate this operation.

7. *Check relay conditions against control drawing.*

8. *Plan "inhibiting" steps for "jogging circuit."* This technique is most useful in step sequencing types of circuits. Selectively interrupt the cycle at progressive intervals so that the machine can be "jogged" to each step. When a particular step is properly performed, the next sequence will be permitted to be executed. *Note*: Before proceeding the reader should have a sound understanding of the ladder diagram. Knowledge of function, types of symbols and their meaning, how to read a ladder diagram, and so forth will be required. The ladder diagram is utilized throughout the remainder of this chapter. Familiarity with this diagramming method will aid the reader in troubleshooting not only power but also control circuits. The ladder diagram is used extensively in industry to represent power and control circuits. Its simplicity and ease of understanding can help reduce time spent troubleshooting. Figure 8.4 shows an example of placing jogging or walking selectors in a pneumatic circuit to selectively interrupt the cycle as it progresses.

Elements designated with a P represent air valves; those with RV are pneumatic relay valves; and those with At are limit valves.

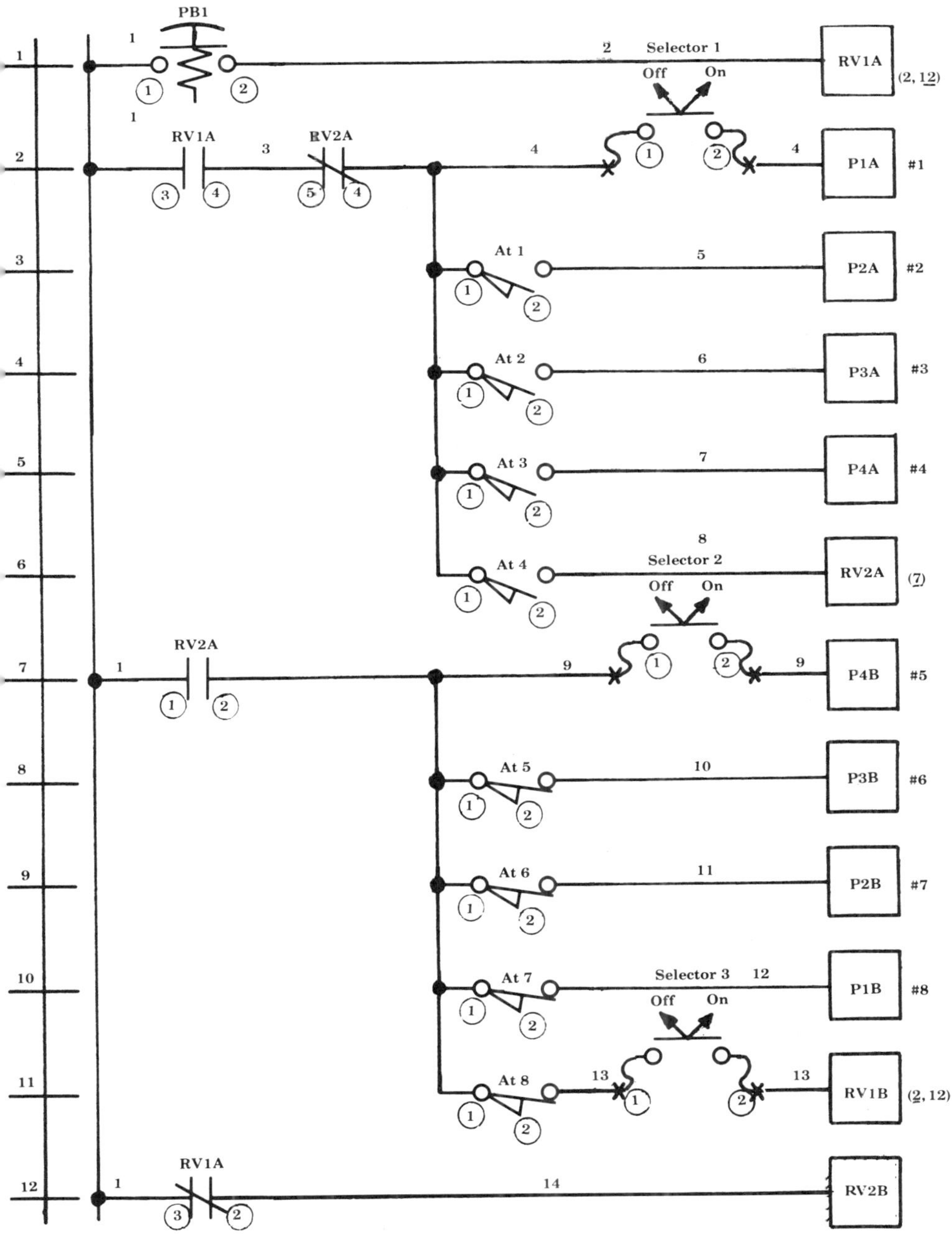

Figure 8.4 "Walking" selectors added to circuit.

Selector 1 was placed to interrupt the path to P1A (power air valve 1 the A pilot). When PB1 is pushed, air will be fed to limits At 1 through 4. If anything moves it will indicate that these limits are not operating correctly. Since the cylinders have been purposely slowed down (step 2 of start-up troubleshooting procedure), it will be easy to isolate any out-of-sequence motions. If nothing moved, selector 1 would be turned to passing and P1A would be pressurized, starting motion 1. When the 4 limit valve is actuated, RV2 will shift. At this point selector 1 could be removed from the circuit and used elsewhere as required because RV2 will be nonpassing in reference line 2.

Nothing should move as long as selector 2 is inhibiting the next operation, step 5. If anything moves, it indicates that one of the limit valves, At 5 through 8, is incorrect or is stuck. If everything is correct, selector 2 would be made passing to allow step 5 to occur.

When limit valve 8 is made at step 8, selector 3 will inhibit the reset of RV1 and the circuit will stop. Selector 2 could not be removed from the circuit at this point because there is still air in it. If the connections of selector 2 were removed with the air on, pressure would drop throughout the entire circuit. Because this circuit uses only detented valves, removing a tube with air on it would not cause any harm. However, if spring-returned power valves or relays were used in the circuit, the pressure drop caused by removal of a tube could cause them to shift back from an actuated position. Never remove a tube or line unless you are sure of the consequence of any air loss and pressure drop and no one is in danger of being struck by a moving machine part.

When selector 3 is made RV1 will reset, causing RV2 to reset immediately after RV1. At this point both are reset from the circuit. With the inhibiting plan complete, the next step can be initiated. The above procedure will help in isolating the malfunctioning leg or component.

9. *Select the simplest automatic cycle and walk through the sequence.* Implement the inhibiting plan as conceived above. The circuit shown above has only one cycling mode shown, so the simplest automatic cycle is the only automatic cycle. In more complex machines the selector would be capable of selecting a "single" cycle or a "repeat" cycle. A semiautomatic cycle would

be considered before a fully automatic cycle if it were simpler. Judgment is important. An hour of careful planning will save 10 hours on the shop floor.

10. *Remove the "walking" device and "run" through the cycle/cycles.* Step 9 proves the hookup and basic sequencing ability of the circuit. Now run several cycles to uncover any mechanical problems.

11. *Proceed to a more complicated cycle.* Repeat the above step as required for alternate modes.

12. *Check emergency procedures.* Depending on the complexity of the circuit, some or all of the above steps will be required. Check the emergency control in every possible operating condition.

13. *Adjust the speed and check all cycles.* Incrementally increase the speed of power devices toward their operating levels. The above steps check the circuit for correct tubing and sequencing ability. This process uncovers any response problems that may develop. Correct as required.

14. *Check shutdown procedure.* This check should be run at the operating speeds encountered on the floor.

15. *Check start-up procedures.* This check is made after the shutdown procedure mentioned above. Often some minor problem overlooked in the design becomes a major problem in reality. Perform the check in that condition which closely approximates the expected operating condition.

When the above procedures have been completed, the machine should be running to the best of its mechanical ability and the troubleshooter's mental ability. Figure 8.5 shows how "walking" selectors might be used in these circuits.

Selector 1 will be used to check the condition of RV2, 3, and 4 as well as limit valves At 1 though 4. If anything moves before selector 1 is made passing, it will indicate a malfunction. At this point only RV1 will be shifted, indicating that the push button has been actuated.

When selector 1 is made passing, action 2 should occur followed by action 3, and so forth. Selector 2 will inhibit the actuation of RV3 until it is evident that the tubing from reference line 9 through 12 is correct. When selector 2 is made passing, action 5 should occur, followed in sequence by 6, 7, and 8.

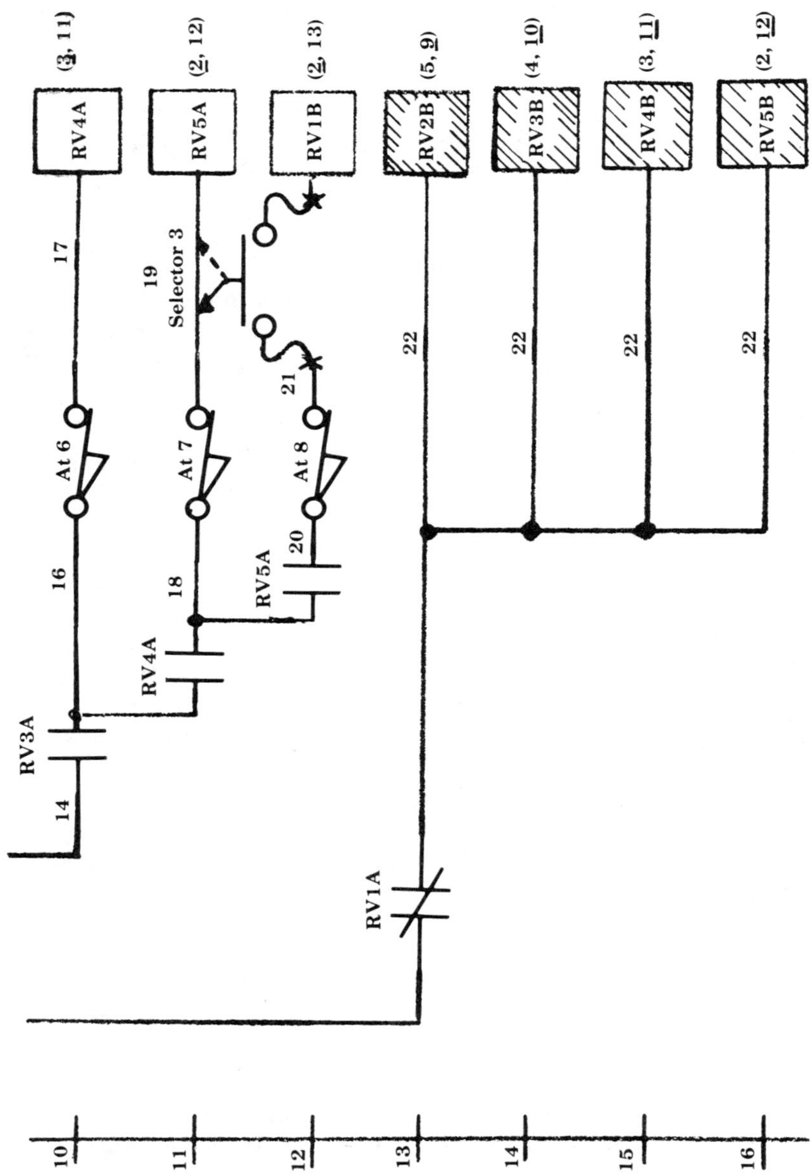
RV4A
(3, 11)
RV5A
(2, 12)
RV1B
(2, 13)
RV2B
(5, 9)
RV3B
(4, 10)
RV4B
(3, 11)
RV5B
(2, 12)
17
19
Selector 3
21
22
22
22
22
At 6
At 7
At 8
16
18
20
RV5A
RV4A
RV3A
14
RV1A
10
11
12
13
14
15
16

Figure 8.5

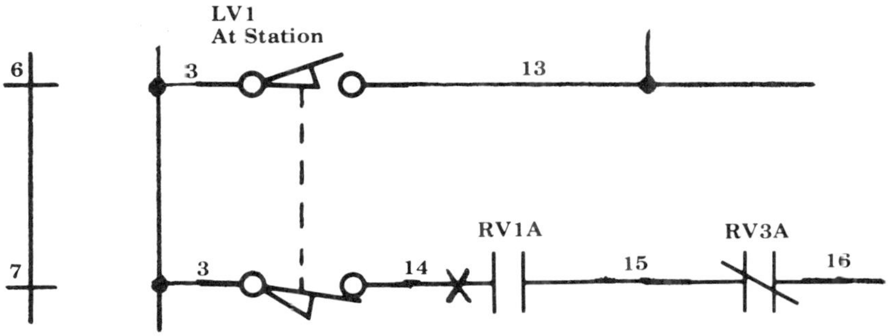

Figure 8.6 "Walking" selectors added to circuit to isolate the index.

Selector 8 will permit a scan of the circuit condition before the circuit is reset.

1. Figure 8.6 shows walking selectors added to the circuit to isolate an index function.
2. Walking selector(s) may be added to interrupt a sequence relay (Fig. 8.7).
3. Another walking technique requires a time delay set for maximum time with its output plugged (Fig. 8.8). The pneumatic timer as shown will ensure that the time delay lasts long enough to allow inspection of the circuit before the action connected with the timer takes place. This will allow enough time to check the slide "out" motion fully before checking the load "down" motion.
4. Figure 8.9 illustrates a walking selector placed to inhibit a master relay reset from being actuated (Fig. 8.9). Placing the selector here again permits the troubleshooter to control events that may happen too fast to follow. The circuit status can be checked before the selector is made passing. Once satisfied that conditions are correct, the selector can be set to permit reset.

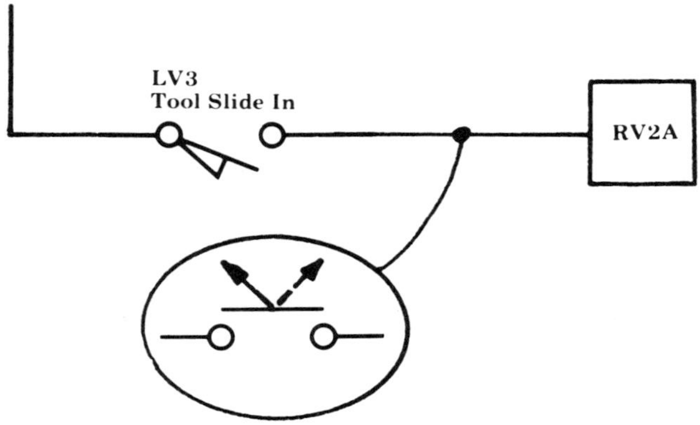

Figure 8.7 "Walking" selectors added to circuit to hold the load "up" and the start "in" to allow time for checking clearances, etc. Also, the position of relay indicator pins can be inspected.

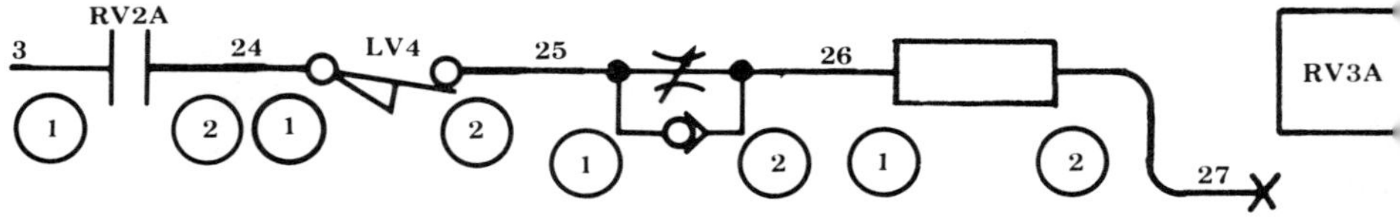

Figure 8.8 Time delay adjusted for minimum flow.

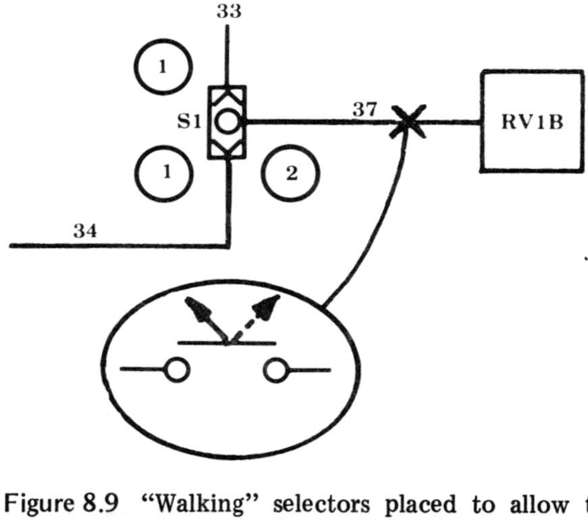

Figure 8.9 "Walking" selectors placed to allow troubleshooter to control events that may happen too fast to follow. The circuit status can be checked before the selector is made passing. Once the troubleshooter is satisfied that conditions are correct, the selector can be set to permit reset.

8.4.2. Troubleshooting a Breakdown

Troubleshooting a breakdown differs from troubleshooting a start-up mainly in the circumstances under which the troubleshooter will be working. At a start-up the circumstances can be controlled to some degree. At a breakdown the circumstances are literally out of control. In addition, there is pressure because a breakdown means downtime, which means loss of production. However, at least the troubleshooter knows that the circuit did work acceptably for some period of time.

Common sense and deductive thinking must control all actions. The following rules will apply in most cases. An accurate control drawing is essential for success. Use the machine manufacturer's maintenance and troubleshooting procedures if possible and if they are available.

1. *Visually inspect the machine.* Become thoroughly familiar with the mechanisms. Ask as many practical questions as possible. Be sure to ask someone who was present at the time of breakdown what the machine did or did not do. Do not get information at second hand.

2. *Check all services to the machine.* Sometimes a breakdown is not a breakdown at all. Is the air on? Is the electricity on? Are all the pressure settings correct?

3. *Determine at what point in its cycle the machine stopped or malfunctioned.* This is the crux of the matter. Generally, properly designed air control systems do not simply fail. More often than not, the mechanical machine fails. However, troubleshooting must be performed on the control system itself to prove that, in fact, the failure was a mechanical one. A common failure is a loose cam that does not properly actuate its limit valve. A machine slide could stick and therefore not allow the next limit in the sequence to be actuated.

Knowing how to read the "status" of the control system will permit you to pinpoint the exact cause. Let us take, for example, a forming machine that has stopped in the middle of a cycle. By asking the appropriate questions it is determined that the machine had been running well and then suddenly stopped. Further questioning reveals that it indexed properly and the load went "up" with no problems. Then the machine stopped. Reference to the sequence of operation and the control diagram indicates that the next step should be:

1. When the load cylinder is fully up it makes LV2.

   A. Tool slide in.

   At this point it is known what should happen next. To determine why the next step did not occur, scan the control panel to see which relay has shifted. Figure 8.10 illustrates this procedure.

2. Figure 8.11 illustrates the circuit as the troubleshooter visualizes it.

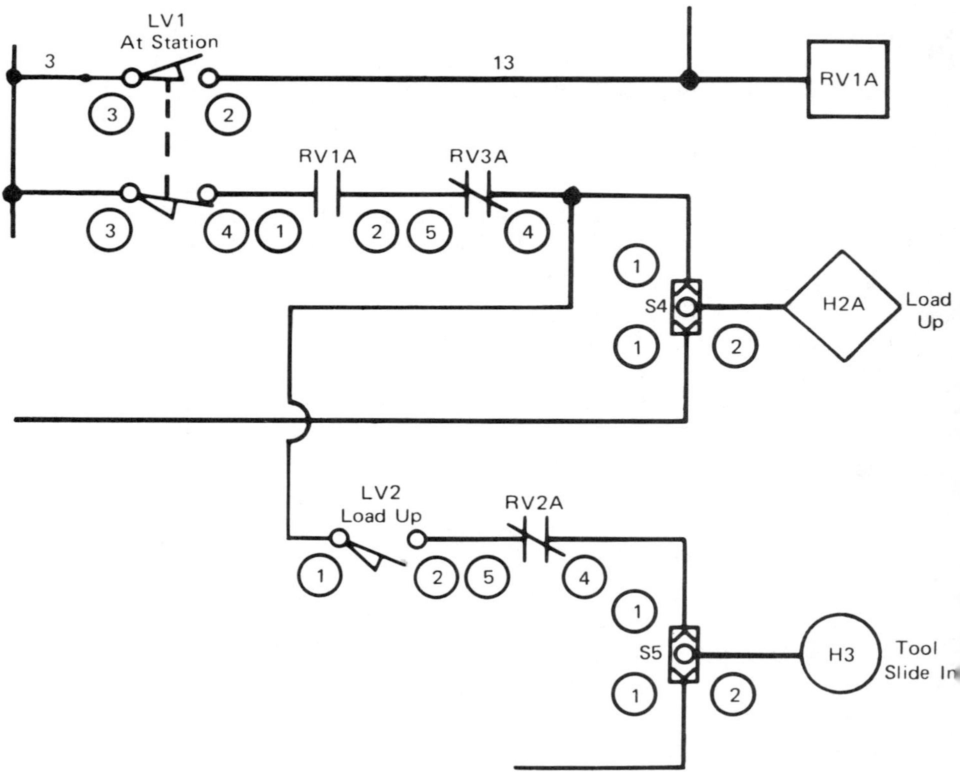

Figure 8.10 Circuit portion as it appears on the print.

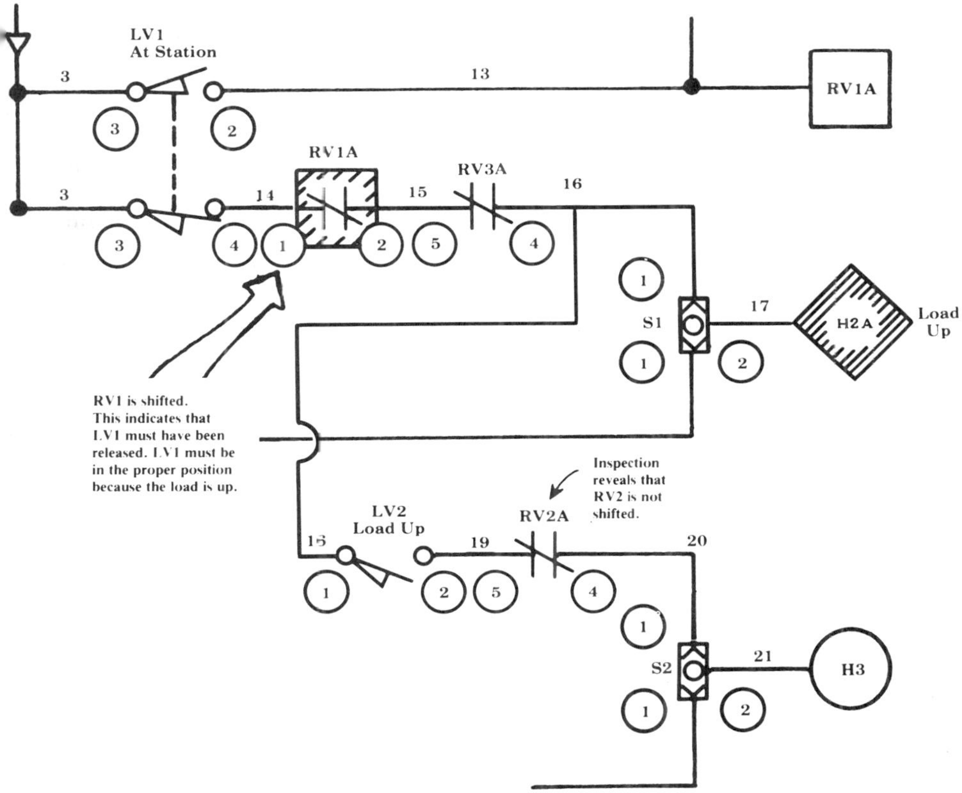

Figure 8.11 Circuit as troubleshooter visualizes it from pin indicator information.

3. It would appear that LV2 is nonpassing, because if it were passing, V2 would allow H3 to be pressurized.
4. An examination of LV2 reveals that the actuator arm has been broken off. When the load cylinder was fully up, its cam could not actuate LV2. The control circuit itself will provide most of the clues required to solve the problem.

Troubleshooting is as much an art as it is a science. As in any art, there is no substitute for common sense and experience. This chapter by no means encompasses all the problems that will be encountered in troubleshooting situations, but it provides a good indication of the direction that must be taken to achieve optimum results.

As was the intent with this entire manual, it is hoped that the approaches and techniques presented here, when taken together, will be of benefit as a basis for further exploration of the application of pneumatics.

9

Current Developments in Pneumatic Directional Control Valves

9.1. INTRODUCTION

No text on pneumatic directional control valves and circuits, or any fluid power topic, would be complete without a chapter highlighting innovations in the area being addressed by the text. That is the purpose of this chapter.

Because pneumatics (the technology of compressed gas used as a power and/or control medium) possesses special and unique advantages (i.e., easy to store and control energy, clean and available at most industrial sites), it has not been displaced by other power or control technologies.

However, engineers and technicians involved in power transmission technology have come to accept and rely on the ever growing role being played by the all-electronic "brain" or control system. This is reinforced by the increasing usage of the microprocessor to control the machine.

As with all innovations or advances in technology, there have been some problems. One of the major problems was one of designing control systems to function at the much lower level control signals now being generated to activate large "power" valves. To accomplish this adjustment to lower power level control signals, innovation has taken the form of miniaturization.

9.2. MINIATURIZATION

In the past, electronic and pneumatic components were applied in much the same way. That is, each component, whether a tube or valve, resistor, or flow control valve, was treated individually. Each was mounted on a chassis and then wired separately or individually. This approach resulted in large groupings or assemblies. This continued until the concept of integrated circuitry spurred the development of new products. In the pneumatic realm, multiple add-on manifolds allowing compact "packaging" became available, as did "sandwich" flow and pressure control devices.

As a natural result of the acceptance of all these devices, the electronic and pneumatic industries started on the road to the development and widespread application of integrated circuits and miniaturization.

9.2.1. Pneumatic Design Considerations

In order to aid the industrial control engineer in interfacing the new "electronic brain," and the time-proven power pneumatics, fluid power component manufacturers analyzed the new type of pneumatic systems required to meet this challenge.

As a result of this analysis, the following pneumatic valve design considerations were developed:

1. Low-power solenoids requiring less than 1.5 W for actuation (actuation by transistor-transistor logic signals), allowing more direct linking of the microprocessor and the valve.
2. Designs that afford ease of maintenance and of troubleshooting.
3. Designs capable of operating at typical industrial pneumatic system pressures, to 116 psig (7 bars).

4. Valve flow capacities (C_v) sufficient to directly operate miniature air cylinders or to be used as the pilot for larger valves.

9.2.2. One Approach

One approach to meeting the above design considerations centers around the electronic industry's standard bus card (4.5 × 6.5 in.). Other design approaches have also been developed, especially in Europe, but the approach shown here is, in our opinion, one of the most practical solutions (Fig. 9.1).

The first step in adapting the printed circuit (PC) board to function as the electropneumatic interface was to develop a pneu-

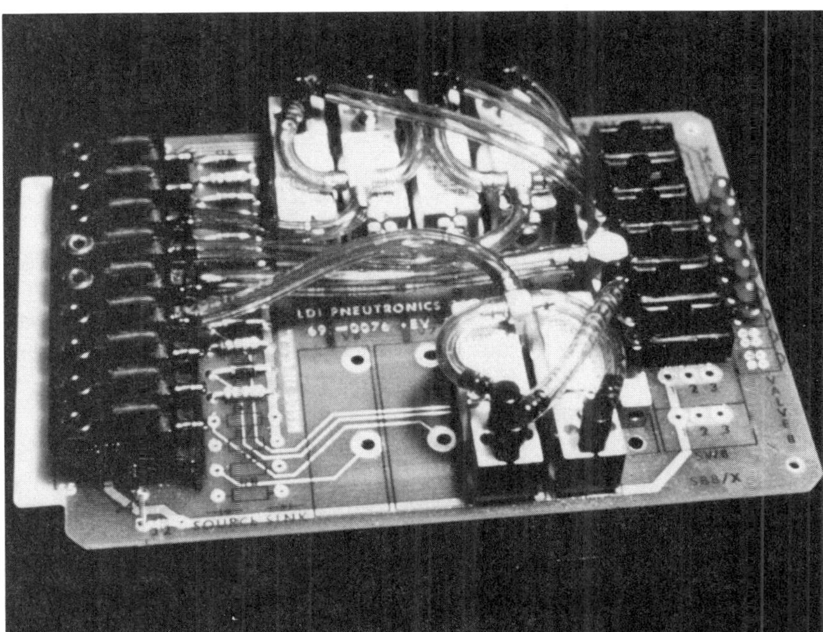

Figure 9.1 (All figures in this chapter courtesy of LDI Pneutronic Corporation, East Pepperell, Massachusetts.)

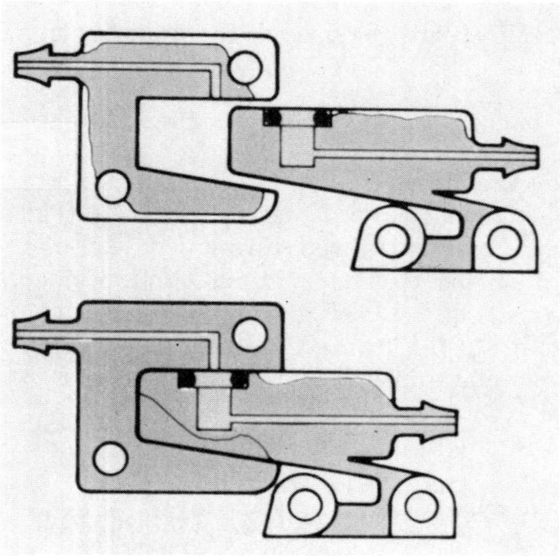

Figure 9.2 Pneumatic edge connectors.

matic edge connector. The industrial style (threaded) connectors were rejected because of problems presented by their inherent design; that is, the forces that are developed by the system pressure are in the same direction as the connector separation direction. This means that the air pressure tries to disconnect, or "blow off," the connector.

This problem was overcome through the development of a new pneumatic edge connector (Fig. 9.2). The new design resulted in the separation and force planes perpendicular or at 90° to each other. This virtually eliminates the "blowoff" problem yet allows for easy installation (or disassembly when required). Sealing between the connector halves is accomplished through the connector's wedge shape and recessed O-ring seal. Now a series of these new connectors are mounted on the edge of the PC board. It is this device that provides a path for the pneumatic signals without a mechanical (threaded) type of connector.

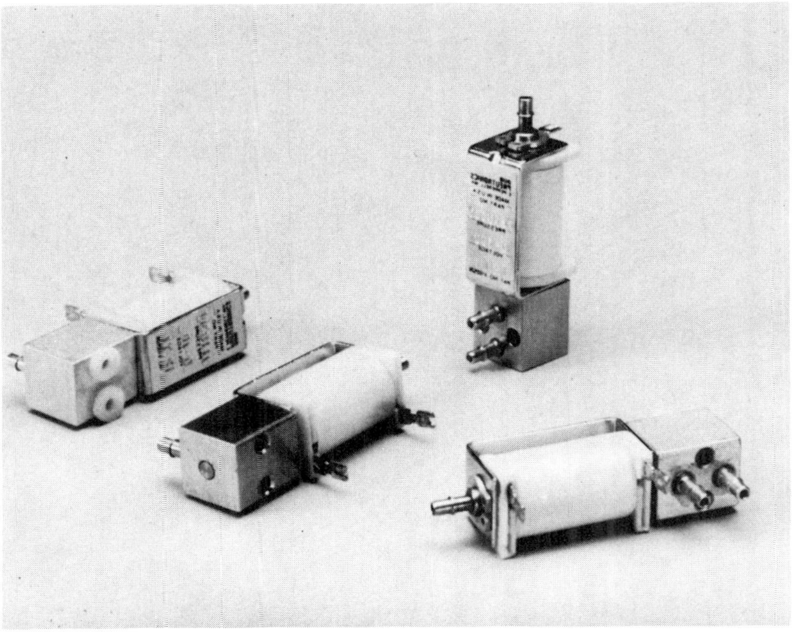

Figure 9.3

In addition to the above connector, new two- and three-way miniature pneumatic valves have been designed (Fig. 9.3). These valves have PC style pin connectors replacing the standard wire leads. This modification allows the use of electronic PC board wave soldering techniques. A similar design approach has been applied to other components such as pressure switches (Fig. 9.4), interface cards (Fig. 9.5), and pneumatic "octal" connector assemblies (Fig. 9.6). One major advantage of this approach is a large number of valves that can be packaged in a small space. As many as 180 valves completely wired and plumbed can fit into 1 cubic foot.

Another benefit of this approach is the simplified troubleshooting techniques employed. Cards can be interchanged between positions or, if a card is found to be defective, detection of the

Figure 9.4

Figure 9.5

Figure 9.6

card and replacement can be accomplished in a matter of seconds, thus returning the machine to production quickly.

9.3. CIRCUIT APPLICATIONS

Pressure control can be achieved utilizing these printed circuit boards, mounted miniature air valves, microprocessor control and a pressure feedback device. A typical circuit for achieving pressure control is shown in Figure 9.7. The output pressure is controlled by momentarily pulsing either the supply or the exhaust valve. A series of these circuits may be installed on a PC board. Each circuit operates in a similar manner.

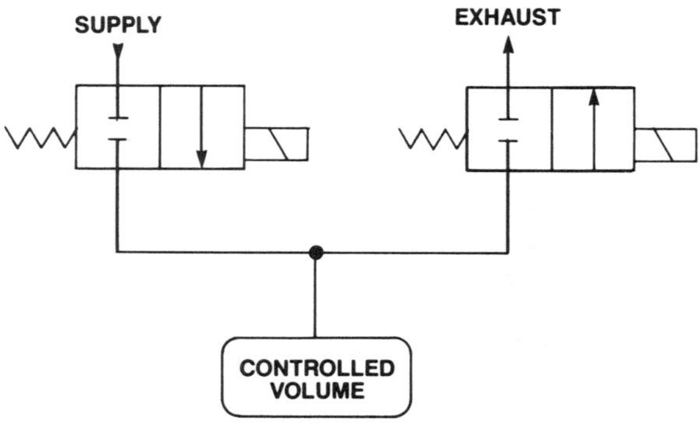

Figure 9.7

Supply pressure would be connected to the inlet of valve A (two-way, two-position) while the outlet of valve B is connected to exhaust (atmosphere on exhaust manifold). When valve A is pulsed, output pressure increases. Likewise, when valve B is pulsed the pressure is decreased.

Implementing the microprocessor control and pressure feedback transducer in this circuit will close the pressure control loop. In this closed-loop system, the microprocessor control will receive the feedback transducer signal, compare it to a desired set point, and signal (pulse) the appropriate valve. By manipulating the microprocessor control program, the duration of the pulse (a form of pulse width modulation) can be changed (shortened or lengthened) depending on the magnitude and direction of any error signal. A simplified flow diagram of the above control loop is shown in Figure 9.8.

Of primary importance to the designer (control and/or systems) is the response time of the system. The valves are, in effect, an orifice. This being the case, it will take some (very small) amount of time to fill and compress or vent and decompress the volume under pressure control. Therefore, the system response time will be directly proportional to the size of the volume under pressure control. It is this capacitance that is the major design factor affecting response times, not valve actuation (response) time and controller logic lag.

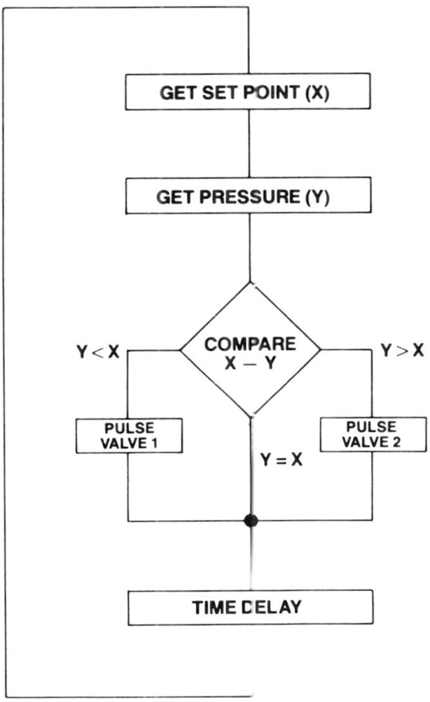

Figure 9.8

Now let us look at three applications of the above pressure control circuit.

Example 9.1 Independent microprocessor pressure control is required for both ends of a small pneumatic cylinder. Flow requirements are minimal, for this is basically a static condition.

Solution This system requires two of the pressure control circuits (Fig. 9.9). The outputs of each control circuit are connected directly to each end of the small pneumatic cylinder. Also connected to the two cylinder ports are the feedback pressure transducers. These signals are then fed back to the controller for comparison with the desired set point. In fact, this circuit will

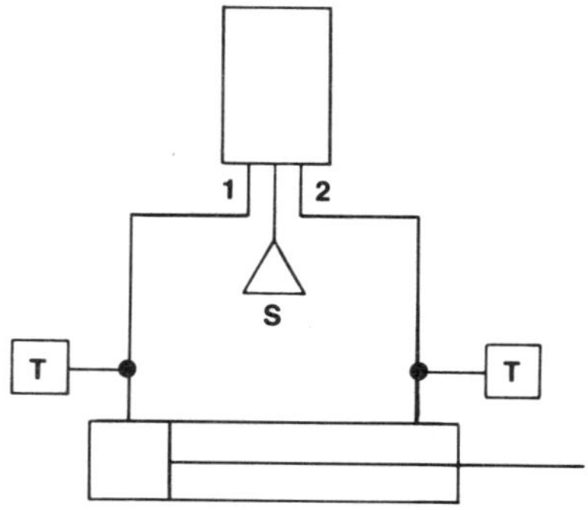

Figure 9.9

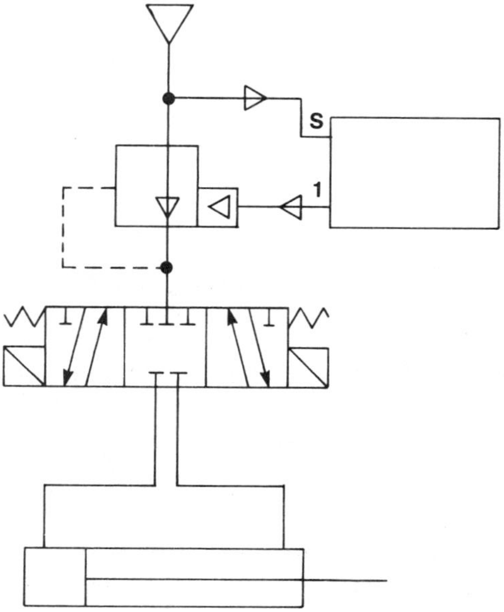

Figure 9.10

achieve independent pressure adjustment of each end of the cylinder.

Example 9.2 Electronic adjustment of the pressure in a clamping cylinder (8-in. bore) is required.

Solution Due to the large air flow requirement, direct cyliner operation is prohibited. Therefore, a pilot-controlled, in-line pressure regulator is needed (Fig. 9.10). The inlet to the pressure control circuit is from the main system. The outlet of this circuit will be used to control the pilot-controlled regulator. As with Example 9.1, a pressure transducer could be added to close the control loop to a microprocessor controller. However, for this circuit to perform up to expectations, the pilot-controlled regulator must have minimum internal friction ("striction"). Even extra volume chambers are not effective when the regulator exhibits high internal friction. Utilizing this approach, pressure control of +0.1 psi over a range of 0–100 psi has been achieved.

Example 9.3 Independent position control of three valves is needed (electronic signals). The speed of actuation of these spring-loaded (normally closed) valves is very slow (Fig. 9.11).

Solution The actuation speed of these valves is slow enough (low flow) to allow direct connection (actuation) between the miniature PC board, mounted valves, and the much larger spring-loaded (off-set) valves. In order to achieve a given valve position (spring-loaded valve) the return spring must be compressed (displaced). To do this a force must be applied to the internal control piston.

Force is equal to pressure times area. In this case, the area is fixed. To achieve variable positioning of the control piston, a variable force is required. With a fixed area, then, we must vary the pressure. In short, this means that the valve position (control piston position) is directly proportional to the control pressure being applied to it.

In this application, three pressure control circuits are needed (one for each spring-loaded valve). The valve position will be fed back to the controller by a potentiometer or similar device. The controller will pulse the correct valves to position the spring-loaded valves. Pressures to system supply pressure could be used

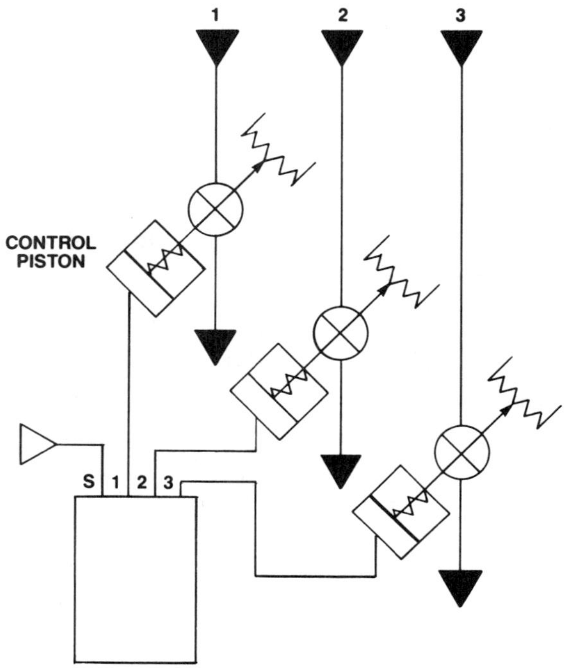

Figure 9.11

should the valves (spring-loaded) encounter increased or unexpected resistance.

The three examples above were given to emphasize that through the combination of electronics and pneumatics their applications are limited only by one's imagination.

Appendix A

Using Flow Coefficients to Design Pneumatic Systems

The common method of sizing pneumatic valves and plumbing is to base them on the cylinder or pneumatic tool, port size. This requires checking the system's ability to supply air to the cylinder at the required pressure. If pressure losses in the lines are too great, it becomes a trial-and-error process to enlarge the pipe sizes and other components. To avoid excessive pressure drops, designers often oversize components.

Cylinders are usually ported for a pipe size large enough to ensure currently desired maximum rod speeds and thrusts. When the cylinder porting is established, a designer may follow suit by using the same size porting for the directional control valves, filters, regulators, speed control valves, and plumbing (see Fig. A.1). The result is often an oversized and an unnecessarily expensive pneumatic system. Use of flow coefficients gives an equally efficient circuit at greatly reduced cost.

Reprinted from Gil Riske, Senior Engineer, Crown Products Division, Hannifin Company, Division of Parker Hannifin Corporation. Using flow coefficients to design pneumatic systems, Bulletin 0601-R3, *Hydraulics and Pneumatics 13*(10): 1–7 (October, 1960).

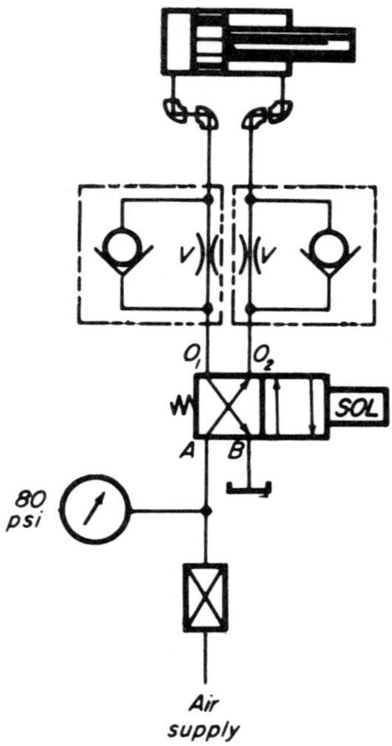

Figure A.1 Typical cylinder circuit with four-way valve control and speed regulation in both directions.

FLOW COEFFICIENTS

Flow coefficient, C_v is a constant which relates fluid flow through a component to pressure drop across it:

$$C_v = \frac{Q}{22.67} \sqrt{\frac{GT}{(P_1 - P_2)P_2}} \tag{A.1}$$

where:

Q = flow of gas in cubic feet per minute (14.7 psi at 60° F)
P_1 = inlet absolute pressure (gauge + 14.7)
P_2 = outlet absolute pressure (gauge + 14.7)
Note: P_2 must be greater than $0.53P_1$

G = specific gravity of flowing medium (air: $G = 1$)
T = absolute temperature of air (460 + °F)

The flow coefficient for a component applies to only one flow path through it. If more than one path is available, the coefficeints vary with each path. The same path may have a different flow coefficient in each direction unless it is symmetrical in both ways. In pneumatic components, a difference of 10% may exist in the maximum flow rates for each direction through the same path.

The use of flow coefficients largely eliminates random guessing and choosing of component sizes. It enables a designer to select the proper component port for each particular application.

FLOW COEFFICIENT CHARTS

Air cylinders are sized for a specified thrust and speed. But directional valves and speed control valves also affect cylinder performance, so you must consider these in working out a final design. When the cylinder bore is determined from required thrust and available air pressure, the charts of C_v values can be used to determine valve size (see Fig. A.2, pages 223–226). Locate on the corresponding cylinder bore chart the intersect of the cylinder's rod speed and thrust. The curves to the right of this point indicate flow coefficients for the supply line and valving. The horizontal lines and above the point represent C_v values for the exhaust system. These are minimum values.

Exhaust system flow coefficients are directly proportional to the distance above the chart zero-speed line. System supply flow coefficients are directly proportional to the spacing between two adjacent vertical lines. The next step is to select directional and speed control valves with combined flow coefficients equal to or greater than the minimum values suggested by the chart.

The combined flow coefficients can be separated into valve and speed control flow coefficients by using the nomogram (Fig. A.3). Locate the combined coefficient on scale B or E and pivot a straight edge at this point to read corresponding combinations of valve and speed valve ratings on scales A and C, or scales D and F.

Only the valves can be sized differently than shown. The original pipe diameter should be maintained even though it means the use of bushings and reducers at the valves. Exception may be made with the use of close nipples at the component ports as well as to connect a directional and speed control valve directly. Longer lengths of the reduced-diameter pipe may adversely affect the capacity of the pneumatic system.

AIR SUPPLY PRESSURE

Different applications obviously require different supply pressures. The accompanying charts are based on pressure of 80 psi at the directional valve inlet (Fig. A.2). Since the speed control valve is usually designed with a spring-loaded check in the free flow direction, the actual inlet pressure at the directional valve should be slightly higher to compensate for the pressure drop created by the spring. Depending on the design, speed control valves may have up to 10 psi pressure drop in the free flow direction. Speed control valves should be rated with flow coefficients for two conditions: (1) the wide open setting in the restricted flow direction and (2) the free flow direction with the metering valve fully closed and the check valve spring removed.

The charts were prepared for supply pressure directed to the cylinder head end. However, they may also be used where the air supplies the rod ends. The flow passages in the cylinder heads are considered as full pipe cross section area minimum. The cylinder length does not exceed 2 ft so as to limit the length of the pipe from the control valve. The pipe is installed exactly as illustrated in the nominal diameter indicated.

THRUST FOR OTHER SUPPLY PRESSURES

Where air supply pressure is other than 80 psi maximum thrust at maximum speed is:

$$T_2 = \frac{(P_s - 25)T_1}{55} \tag{A.2}$$

where:

T_1 = maximum rod thrust (pounds) read on chart at desired maximum speed

T_2 = maximum rod thrust (pounds) at same speed but with supply pressure P_s
P_s = actual air supply pressure (psig)

Maximum speed remains constant because exhaust flow coefficient remains constant.

To use the curved lines of the charts, when changing supply pressure, determine the intersect of the straight and curved lines from Eq. (A.2). Also determine the maximum rod thrust at static condition (zero speed) with the following equation:

$$T_3 = A \times P_s \tag{A.3}$$

where:

T_3 = static rod thrust (pounds)
A = piston area (in.^{2})

CYLINDER CYCLING TIME

After choosing directional and speed control valves to meet thrust and speed requirements, it may be found that reversal time is too great. This occurs at the end of the stroke, when it is necessary to reduce air pressure on one side of the piston to provide the required thrust in the reverse direction.

The time to exhaust the cylinder depends on flow capacity of the components in the exhaust line and the volumes of the pipes and cylinders. Because exhaust time may be considerable, it should be calculated from the following equation:

$$E = \frac{1.432K_1(AS + R)}{C_v K_2} \tag{A.4}$$

where:

E = exhaust time (minutes)
K_1 is found in Table A.1
K_2 is found in Table A.3
A is found in Table A.2
S = rod stroke
C_v = combined exhaust valve system flow coefficient
R = pipe volume from valve to cylinder (ft^3)

Table A.1 Values for K_1[a]

(Exhaust Time Constant) Supply Pressure—psi	K_1
40	0.050
60	0.074
80	0.092
100	0.107
120	0.12
140	0.13
160	0.14

[a]Exhaust time constant.

SELECTING VALVES USING FLOW COEFFICIENTS

Required:

1. Working thrust of 1750 lbs horizontally directed to move 18 in. in 3 s.
2. Return stroke force of 250 lbs to move 18 in. in 1 s.
3. Determine cycle time for this system. Assume valve shift time equal to five electrical cycles.

Given:

Air supply pressure: 80 psig.

Method of Solution:

1. Calculate the minimum cylinder bore for a static thrust of 1750 lbs:

$$\text{Area (in.}^2) = \frac{\text{thrust (lb)}}{\text{supply pressure (psi)}}$$

$$= \frac{1750}{80}$$

$$= 21.9 \text{ in.}^2$$

A 6-in. diameter cylinder which has a head end area of 28.27 in.^{2} is suitable.

Table A.2 Areas of Cylinder Bores

Cylinder Bore —in.	Head End Area —ft.^{2}	Rod End Area —ft.^{2}
1½	0.012	0.010
2	0.022	0.020
2½	0.034	0.032
3½	0.057	0.051
4	0.087	0.082
6	0.196	0.186
8	0.365	0.349
10	0.545	0.528

Table A.3 Correction Factor K_2 for Nominal Pipe Diameters

Combined valve C_V	¼ in.	½ in.	¾ in.	1 in.
1	0.94	1.00	1.00	1.00
2	0.77	0.95	0.98	0.99
3	0.60	0.90	0.96	0.98
4		0.85	0.94	0.97
5		0.80	0.92	0.96
6		0.75	0.90	0.95
7			0.88	0.94
8			0.86	0.93
9			0.84	0.92
10			0.82	0.91
11				0.90
12				0.89
13				0.88
14				0.87

2. Calculate extending speed:

$$\text{Speed} = \frac{18 \text{ in.} \times 60 \text{ s/min}}{3 \text{ s} \times 12 \text{ in./ft}}$$

$$= 30 \text{ ft/min}$$

3. Determine minimum flow coefficient of the combined directional and speed control valves for the cylinder work stroke. Find the point for 1750 lb thrust and 30 ft/min speed on the 6-in. cylinder chart. Interpolating, supply C_v is 1.5 and exhaust C_v is 0.75.

4. Calculate rod end area required for return stroke thrust of 250 lb:

$$\text{Area (in.}^2\text{)} = \frac{250}{80}$$

$$= 3.13 \text{ in.}^2$$

Since the piston rod end area, with a 1 3/8-in.-diameter rod, is 26.79 in.^{2}, the 6-in.-diameter cylinder is suitable.

5. Calculate return stroke speed:

$$\text{Speed} = \frac{18 \text{ in.} \times 60 \text{ s/min}}{1 \text{ s} \times 12 \text{ in./ft}}$$

$$= 90 \text{ ft/min}$$

6. Determine minimum flow coefficient of combined valving for the return stroke. At the intersect of 250 lb thrust and 90 ft/min speed on the 6-in. cylinder chart, minimum exhaust flow coefficient is 2.25 and the minimum supply flow coefficient is 2.0.

7. Select directional and speed control valves with combined flow coefficient in excess of the minimum ratings from steps 3 and 6. Separate the combined flow coefficients by using the nomogram. Locate the combined flow coefficient on scale B or E, which cover different ranges. Pivot a straight edge at this point to read corresponding combination of directional and speed control valve ratings on scales A and C, or scales D and F.

Prepare a table indicating separate and combined flow coefficients of the valves for the various flow paths. Working with data of valves of known flow coefficients, determine by trial and error the combination that meets the minimum capacity ratings for all conditions.

This cylinder should perform to specifications with a four-way 1/2-in. NPT directional valve, a 1/2-in. NPT speed control valve in the line to the head end, and a 3/8-in. NPT speed control valve in the line to the rod end.

8. To calculate cycle time for the 6-in.-diameter, 18-in. stroke cylinder using a valve shift time of five electrical cycles, compute time to energize the valve for cylinder work stroke:

$$\text{Five cycles} \times \frac{1}{60 \text{ cps} \times 60 \text{ s/min}}$$
$$= 0.001 \text{ min}$$

Time to exhaust rod end of cylinder is, from Eq. (A.4)

$$E = \frac{1.732 \times 0.092\ (0.186 \times 1.5 + 0.006)}{0.75 \times 1.0}$$
$$= 0.061 \text{ min}$$

Time for work stroke is

$$\frac{18 \text{ in./stroke}}{12 \text{ in./ft} \times 30 \text{ ft/min}} = 0.050 \text{ min}$$

Time to de-energize the valve for return stroke is the same as that to energize it, 0.001 min. Time to exhaust head end side of cylinder, using Eq. (A.4)

$$E = \frac{1.732 \times 0.092\ (0.196 \times 1.5 + 0.006)}{2.50 \times 0.98}$$
$$= 0.016 \text{ min.}$$

Time for return stroke is

$$\frac{18 \text{ in./stroke}}{12 \text{ in./ft} \times 90 \text{ ft/min}} = 0.017 \text{ min}$$

Minimum time/cycle is total, or 0.146 min. Maximum number of cycles/min:

$$\frac{1}{0.146 \text{ min/cycle}} = 6.85 \text{ cpm}$$

In this example, the total exhaust time is 53% of the cylinder cycle time. To shorten the exhaust part of the cycle, two methods are available:

1. Use a quick exhaust valve.
2. Use a pressure regulator in the cylinder rod-end line. This will conserve compressed air and limit free air volume that must be exhausted.

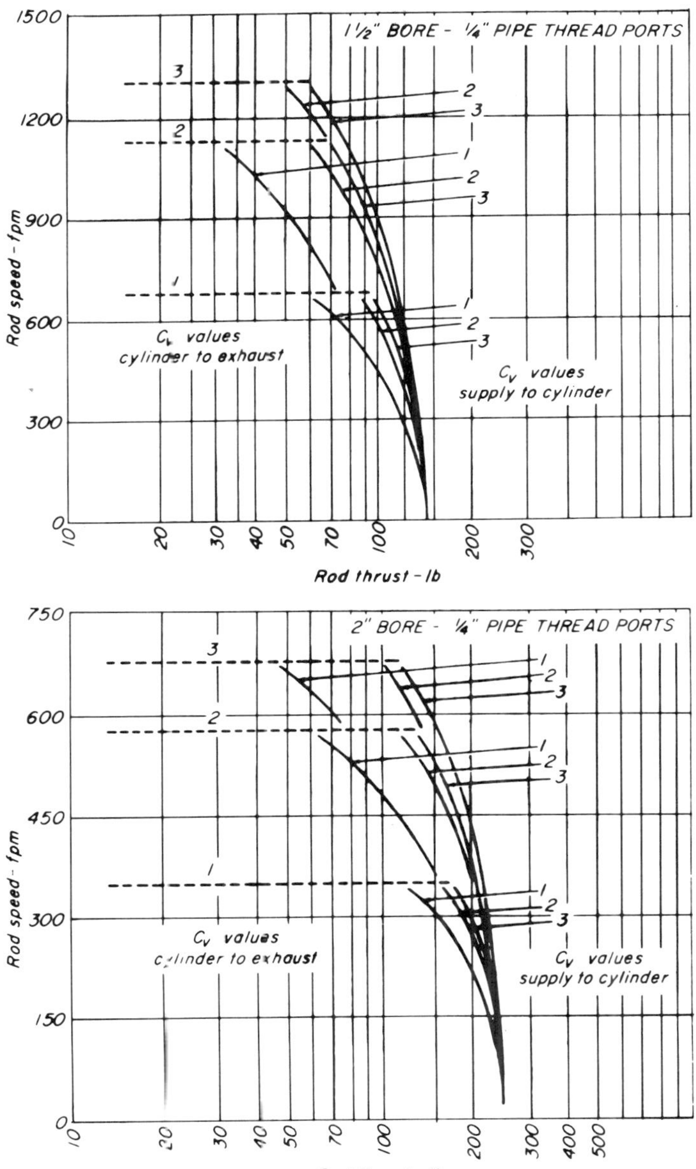
1½" BORE - ¼" PIPE THREAD PORTS
Rod speed - fpm
1500
1200
900
600
300
0
Rod thrust - lb
10
20
30
40
50
70
100
200
300
C_v values cylinder to exhaust
C_v values supply to cylinder
2" BORE - ¼" PIPE THREAD PORTS
750
600
450
300
150
400
500

Figure A.2

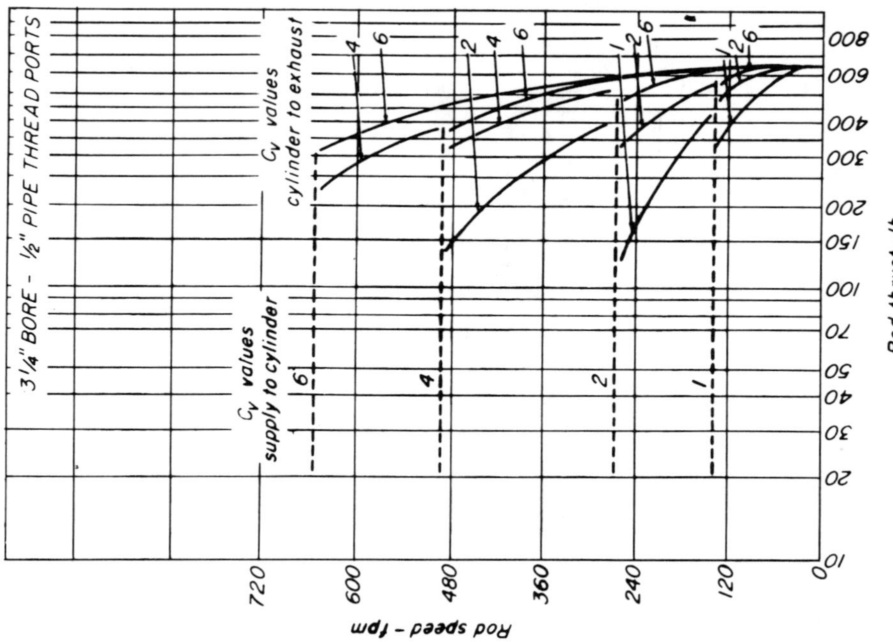
3 1/4" BORE - 1/2" PIPE THREAD PORTS
Cv values cylinder to exhaust
Cv values supply to cylinder
Rod thrust - lb
Rod speed - fpm
10
20
30
40
50
70
100
150
200
300
400
600
800
0
120
240
360
480
600
720
1
2
4
6

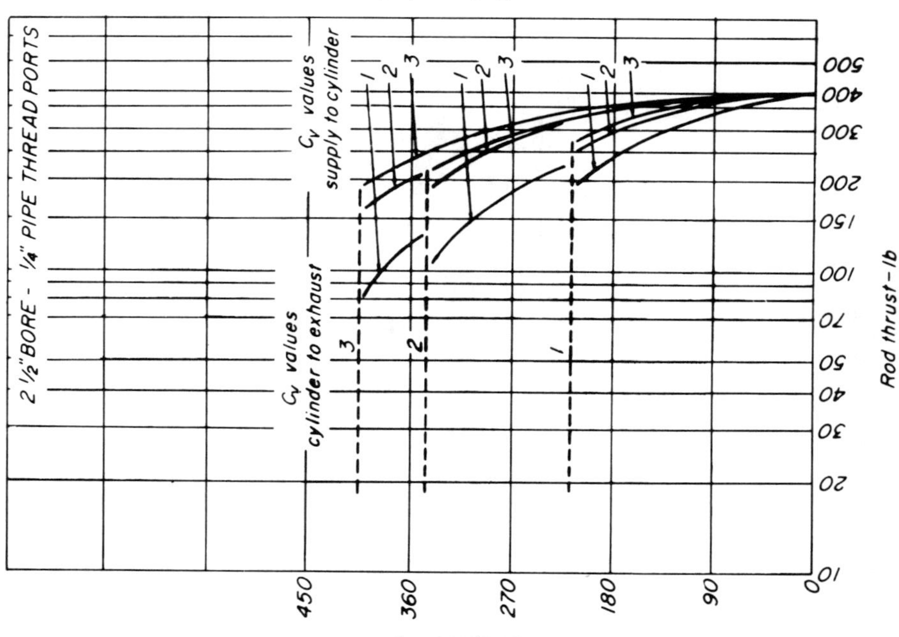
2 1/2" BORE - 1/4" PIPE THREAD PORTS
Cv values supply to cylinder
Cv values cylinder to exhaust
Rod thrust - lb
Rod speed - fpm
10
20
30
40
50
70
100
150
200
300
400
500
0
90
180
270
360
450
1
2
3

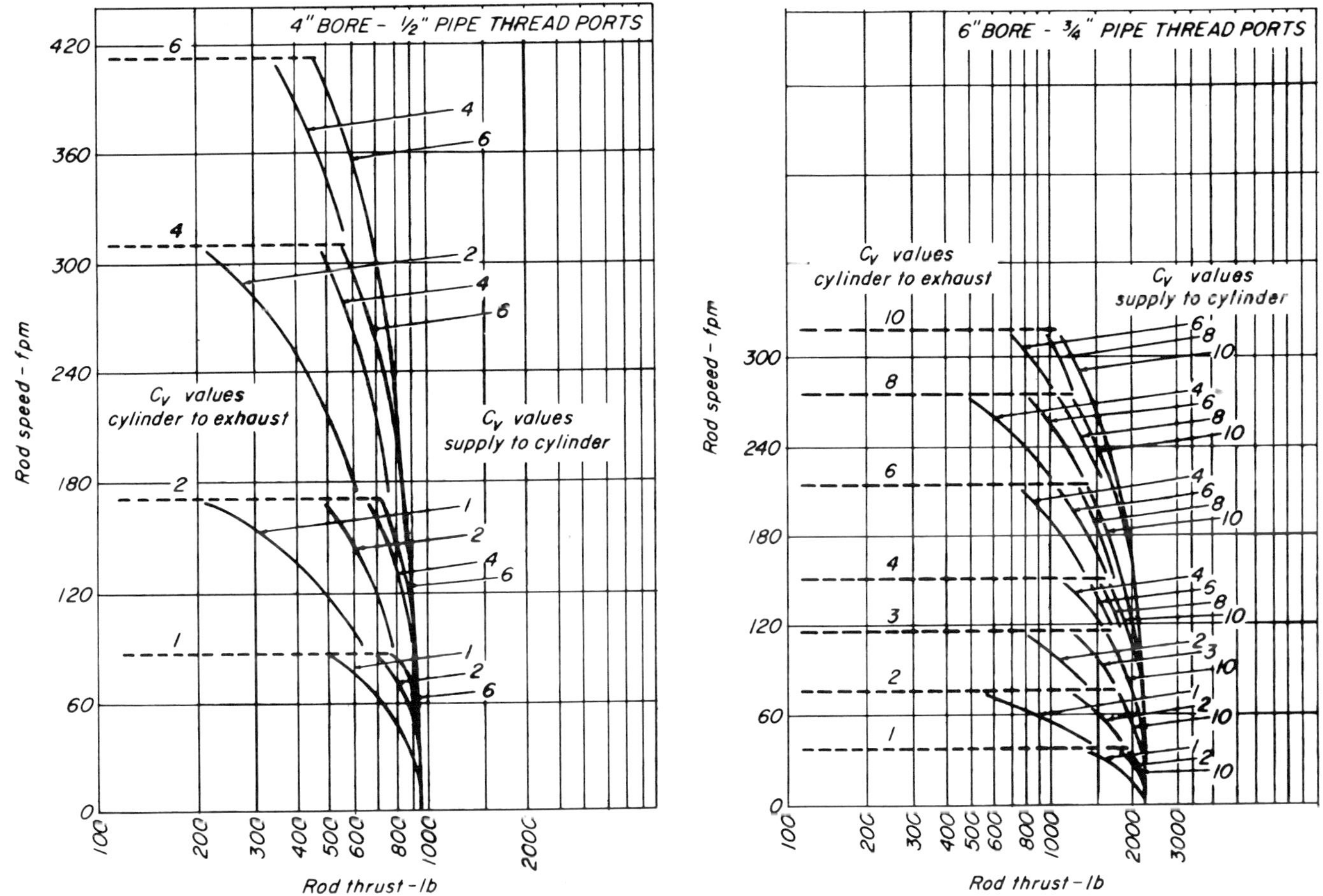

Figure A.2 Continued

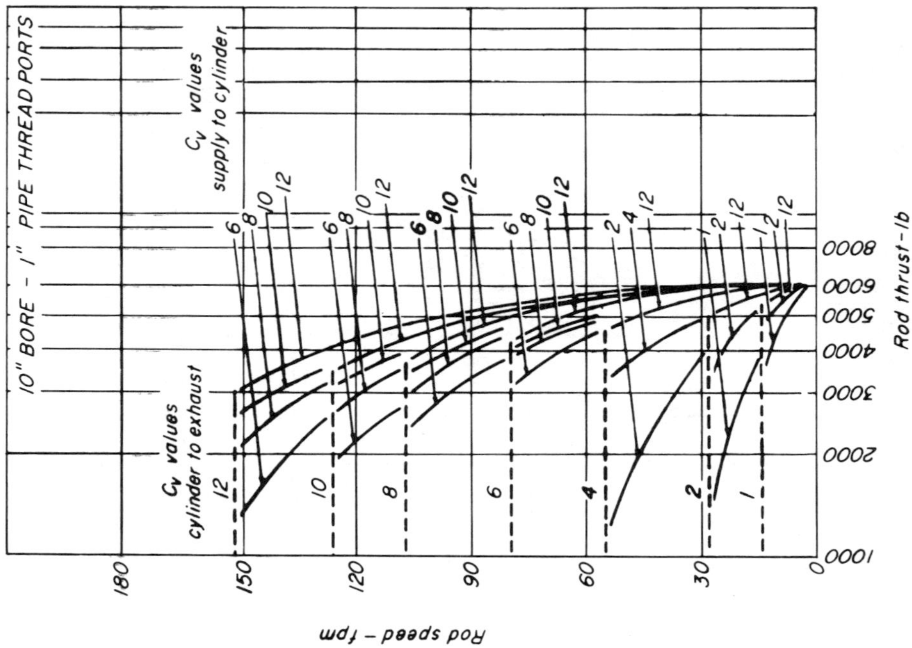
10" BORE - 1" PIPE THREAD PORTS
C_v values supply to cylinder
C_v values cylinder to exhaust
Rod speed – fpm
Rod thrust – lb

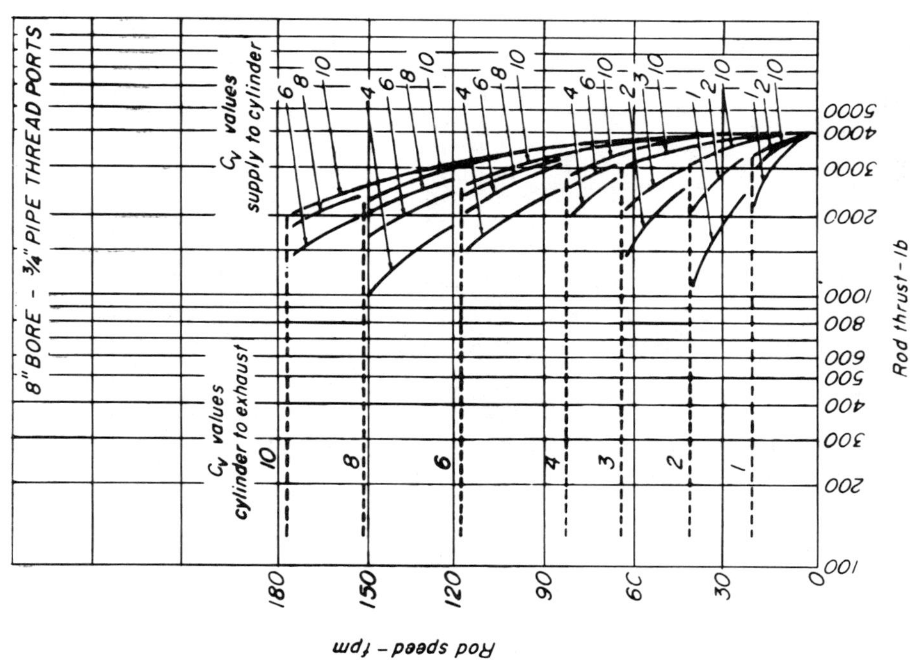
8" BORE - 3/4" PIPE THREAD PORTS
C_v values supply to cylinder
C_v values cylinder to exhaust
Rod speed – fpm
Rod thrust – lb

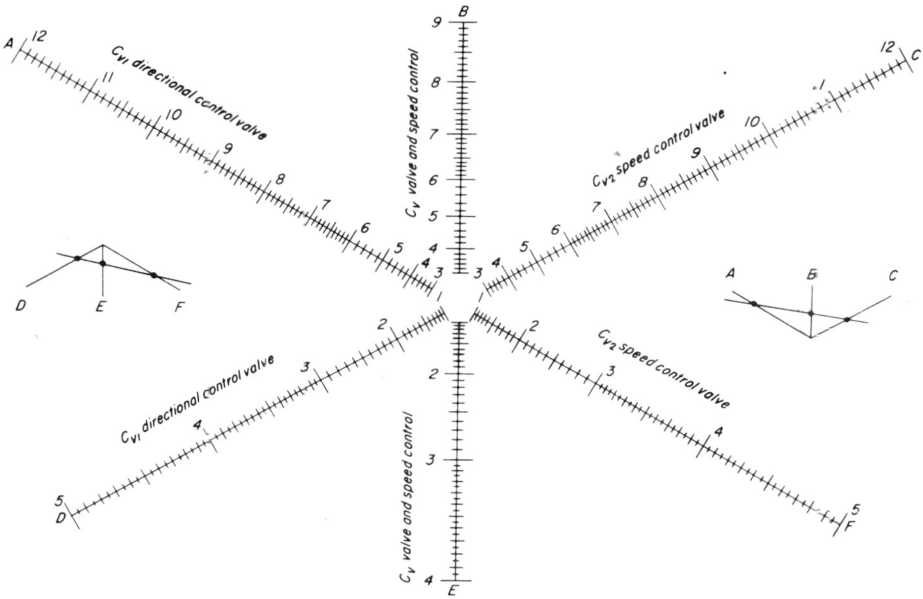

	Nomogram Values			Valve Ratings		
Flow path through valve	Scale E Combined C_V	Scale D Directional valve C_V	Scale F Speed control valve C_V	Directional valve C_V	Speed control valve C_V	
				1/2" NPT	1/2" NPT	3/8" NPT
A to O_1	1.5	1.80	2.63	4.7	4.2[1] to 4.5[3]	—
O_2 to B	2.50	4.6	3.00	4.6	3.00[2] max	—
A to O_2	2.0	4.7	2.25	4.7	—	2.63[1] to 3.82[3]
O_1 to B	0.75	4.6	1	4.6	—	2.44[2] max

Figure A.3 Nomogram to determine valve flow coefficients from combined coefficient. 1—Free flow (Needle closed—No valve spring) 2— Restricted flow through needle valve) 3—Free flow plus full open needle valve—no valve spring.

Appendix B

The Art of Analyzing Pneumatic Circuits

The scope of this appendix will be limited to the preparatory steps in basic circuit analysis of steady flow systems as well as a discussion of some factors affecting the degree of accuracy of the calculations. The intent of an analysis is not to oversize a system purposely, but rather it is to check against unintentional undersizing. It is also intended that this method be applied directly to the pneumatic system on the machine or anywhere in the plant. For this analysis, pressure gauge taps may be made at pipe tees inserted into the lines for this purpose.

The most difficult series of steps in any analysis are concerned with getting started. With pneumatic circuits, this is more often the rule than the exception. Basically, once the flow factors of the system sections have been determined, the relationship between the pressure losses and air flow rate can be evaluated. The difficult phase concerns determining the system section flow factors.

Reprinted from Gil Riske, Senior Engineer, Crown Products Division, Hannifin Company, Division of Parker Hannifin Corporation. The art of analyzing pneumatic circuits. *Proceedings of the National Conference on Industrial Hydraulics*, October 20–21, 1960. Copyright 1960, Illinois Institute of Technology Research Institute, Chicago, Illinois, pp. 238–245.

USING FLOW FACTOR EQUATIONS

Although capacity coefficients are commonly used to rate industrial fluid components, the use of flow factors in pneumatic circuit analysis is suggested because it represents a direct all-around means for getting the job done. The National Bureau of Standards originally recommended the use of flow factors to rate pneumatic valves and other components [1]. The concept of flow factors stems from the classically derived equations which were frequently applied, early this century, to the solution of problems of gas and air flow through pipes. Also, many of the empirical equations developed later included a relationship that expressed the pipe friction as a function of the pipe diameter. Another item favoring the use of the flow factor equations is that they are readily applied to analytical solutions.

Numerically, the flow factor of a pneumatic component may be defined as the maximum air flow rate that can be put through it for each pound per square inch of absolute supply pressure. For example, a valve with a flow factor equal to one would permit a maximum air throughput of 95 standard ft^3/min (scfm) with an absolute supply pressure of 95 psi.

The basic flow factor equation is

$$Q_0 = 1.153F_0(P_1{}^2 - P_2{}^2)^{1/2} \tag{B.1}$$

or

$$Q_0 = 1.153F_0(P_1{}^2 - P_2{}^2)^{1/2} \tag{B.2}$$

and, for other temperatures and specific gravities, the flow factor F_0 can be converted into F_0 as follows

$$F_0(G_0T_0)^{1/2} = F_0(G_0T_0)^{1/2} \tag{B.3}$$

where

Q_0 = air flow rate, standard ft^3/min (68° F and 14.7 psia)
F_0 = flow factor at standard conditions (68° F and 14.7 psia)
T_0 = temperature at standard conditions (460 + 60° F = 528° R)
G_0 = specific gravity of air = 1
Q_0 = air flow rate (ft^3/min) at conditions other than standard
F_0 = flow factor at conditions other than standard

T_0 = temperature at conditions other than standard
G_0 = specific gravity of gas if different from air

Note that $F = Q_{max}/P_1$ when $P_2 = P_1/2$ or less and that when P_2 is less than $P_1/2$, the value $P_1/2$ should be used for P_2.

Equation (B.1) and (B.2) are variations of those proposed by the National Bureau of Standards for rating pneumatic valves. Equation (B.3) includes the correction factors for specific gravity and temperature. Since the constant 1.153 is represented by the critical pressure ratio of $P_2/P_1 = 0.5$ in the term $(P_1^2 - P_2^2)^{1/2}$ (which is also a mean value for all gases) it can be used without correction.

EQUIVALENT FEET OF STRAIGHT PIPE

The National Bureau of Standards suggestion for determining the total flow factor of a series of components has been reworked and simplified to a graphical process whereby the flow factors may be related to lengths of straight pipe or tubing of any internal diameter [2]. This is a real convenience because it makes it possible to relate all fixed restrictions in a pneumatic system in terms of equivalent feet of straight pipe of any common internal diameter. These fixed restrictions may include pipe, pipe fittings, valves, and other air line compoents that exhibit only one pressure loss at controlled conditions; and in the process of determining the system section flow factors, all component flow capacities must be expressed in terms of equivalent feet of straight pipe of a common inernal diameter. After this is done, it is simply a matter of adding these lengths and converting this total length into the section flow factor.

CONVERTING TO FLOW FACTORS

A variety of methods are used to express the fluid flow capacities of pneumatic system components. It is necessary to convert these expressions to flow factors. For unrated components, suggestions are offered for estimating the flow factors with a reasonable degree of accuracy.

Pipe fittings and valves (elbows, tees, globe valves, gate valves, check valves, etc.) are rated in many supply house catalogs in

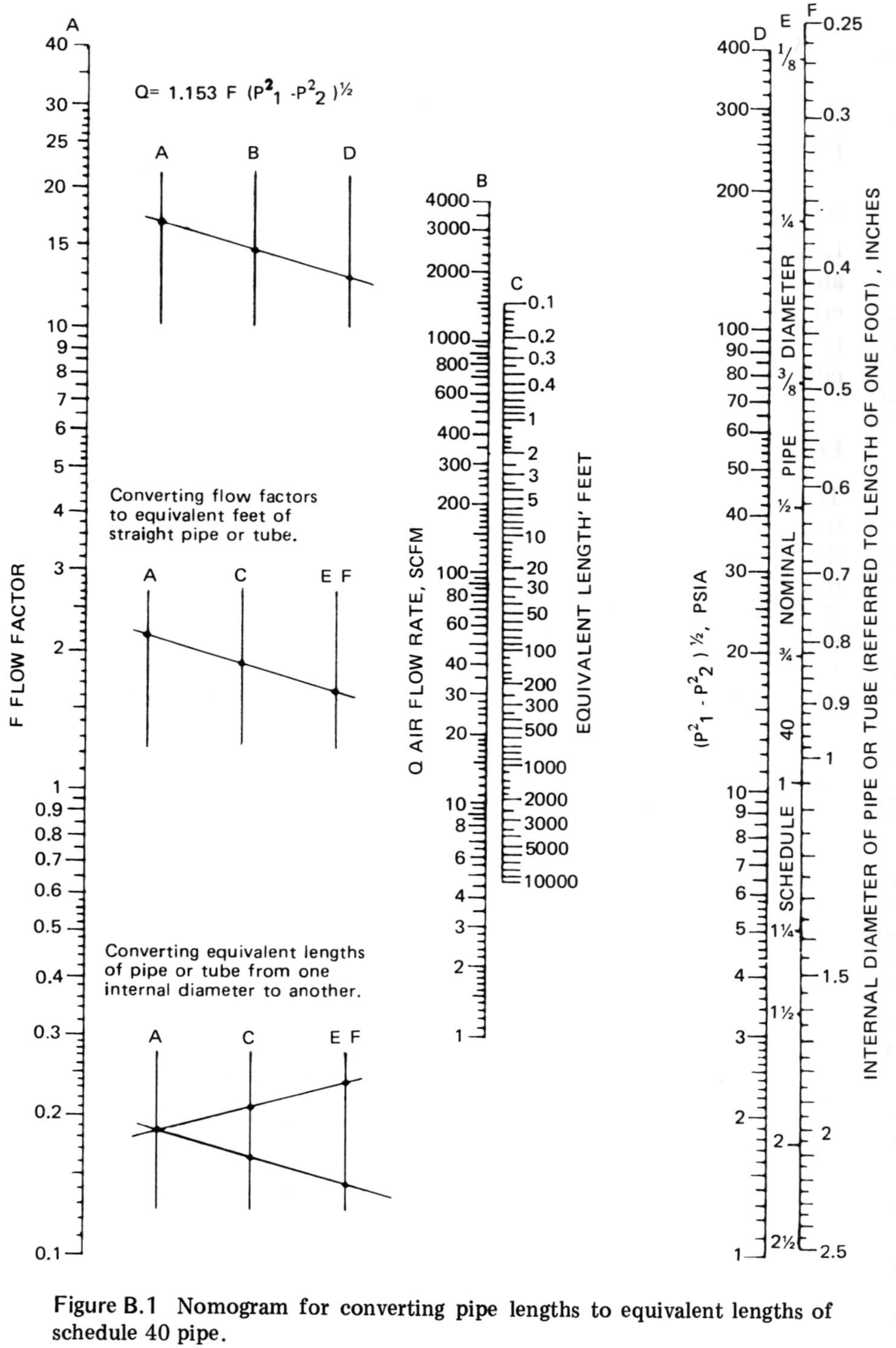

Figure B.1 Nomogram for converting pipe lengths to equivalent lengths of schedule 40 pipe.

equivalent feet of straight pipe (schedule 40). If the pipe to be used does not have the wall thickness of schedule 40, then the actual length may have to be converted to an equivalent length of schedule 40 pipe with the use of Fig. B.1.

Valve manufacturers, on the other hand, often rate their products with flow capacities called capacity coefficients (C_v). A close approximation in converting is to assume that the C_v rating is numerically twice that of the equivalent flow factor at standard conditions. For example, a C_v rating of 3 would be equivalent to a flow factor rating of 1 1/2.

Some manufacturers publish flow data for their products in the form of pressure losses at specific air flow rates and supply pressures. The flow factor may be determined with the use of Eq. (B.1). and Figures B.1 and B.2.

The direct conversion of other air and gas flow coefficients into flow factors will not be attempted here. Instead, it is suggested that flow coefficients other than C_v be used to solve for the upstream and downstream pressures and air flow rates closest to the estimated use, and these data be converted to the flow factor. It is a good idea to tabulate these flow factors for future reference.

If the flow capacity data of any unrated components are not available, try contacting the manufacturer. However, if the component is on hand, it may be more convenient to do the testing or have it tested. If these alternatives are not available, calculate the expected flow factor of the component from the physical dimensions of the passage contours [3].

There are special problems in analyzing a pneumatic system already installed on a machine or in a building. For one thing, the unrated components are often not easily removed from the circuit for testing. Also, flow data may not be available from the manufacturer, or it may be that the item was made especially for the application and its capacity is unknown. Whatever is the case, something has to be done to determine the capacity ratings. One method involves the use of pressure gauges both upstream and downstream from the unrated unit, as well as another pair of gauges both upstream and downstream from a rated section of the system that handles the same volume of air during the cycle. This method involves the use of the following equation:

$$F_a = F_b\,|(P_3{}^2 - P_4{}^2)/(P_1{}^2 - P_2{}^2)|^{1/2} \qquad \text{(B.4)}$$

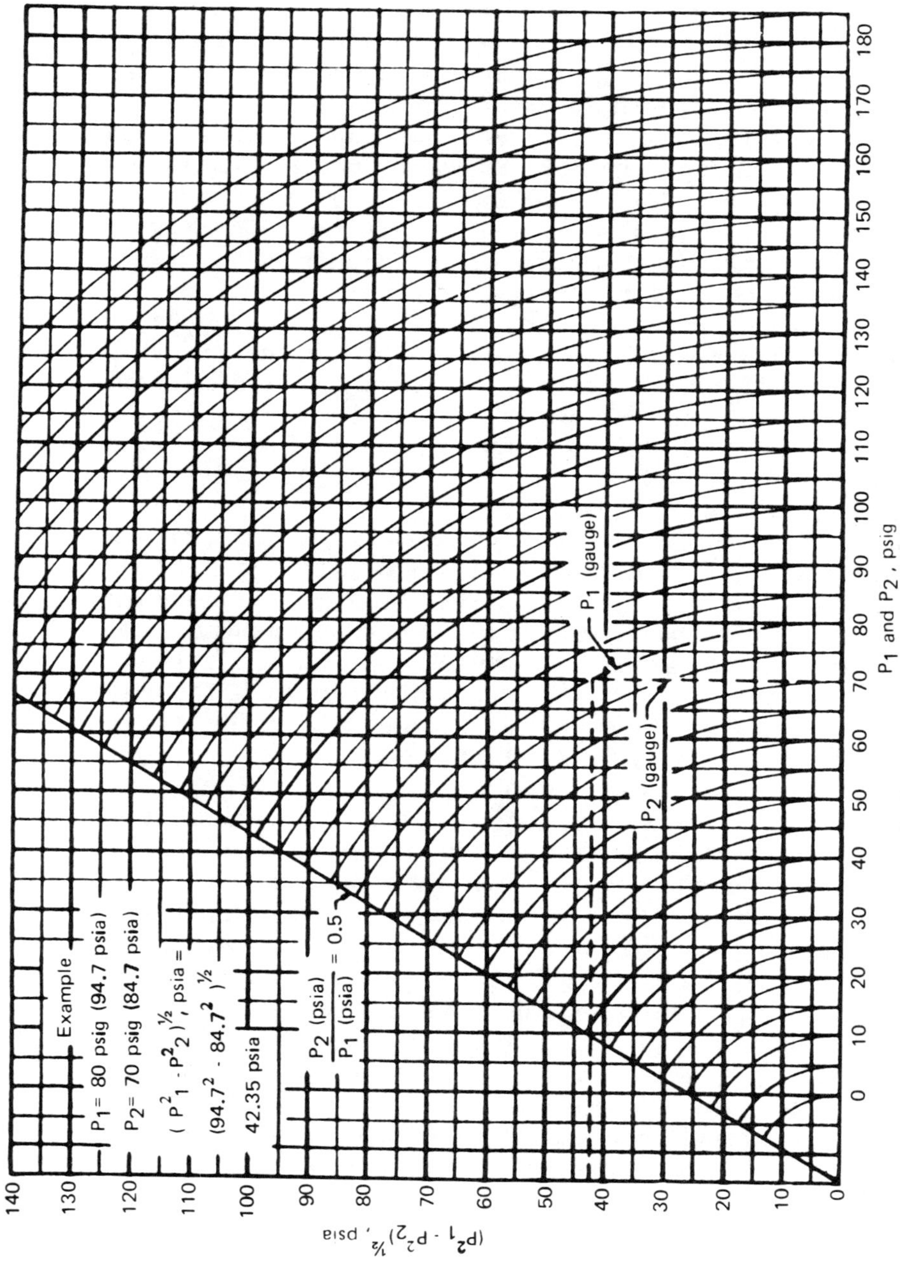

Figure B.2 Graphical computation of $(P_1^2 - P_2^2)$.

where:

F_a = flow factor of the unrated component
F_b = flow factor of the rated component
P_1 = upstream pressure (psia) of component *a*
P_2 = downstream pressure (psia) of component *a*
P_3 = upstream pressure (psia) of component *b*
P_4 = downstream pressure (psia) of component *b*

Equation (B.4) can be solved with the aid of the nomogram in Figure B.1 and the graph in Figure B.2.

A differential pressure gauge is preferred for these tests even though the upstream pressure must still be measured separately. The rated section may be no more than pipes and fittings that provide a pressure loss of 5 to 10% of the upstream pressure at the measured flow conditions. This method ignores the direct measurement of the air flow rate, although it can be determined with the aid of Eqs. (B.1) or (B.2), using the pressure gauge readings and the flow factor of the rated section.

MAKESHIFT FLOW METERS

The lack of a flow meter for an on-the-spot check of a pneumatic circuit or component may be a discouraging experience. However, these devices are not as plentiful as pressure gauges, nor are they as portable or as reasonably priced. Yet, the direct measurement of air flow rate may be desirable even with a makeshift device of reasonable accuracy. Such a device can be made with only pipes, fittings, and pressure gauges arranged as shown in Figure B.3. Be sure to use pipe, tees, and nipples of the same nominal pipe diameter and schedule. Also, the pipe union can be any convenient size equal to or larger than the nipples. A differential pressure gauge is preferred because the pressure loss can be read directly. Also, locate this pipe arrangement upstream from the pneumatic system so that the normal operating cycle is not affected by the additional pressure loss.

While this method of flow measurement is not recommended to replace any of the orthodox flow meters, it will prove invaluable when no standard flow meter is available. However, disadvantages of this method of air flow metering arise from the

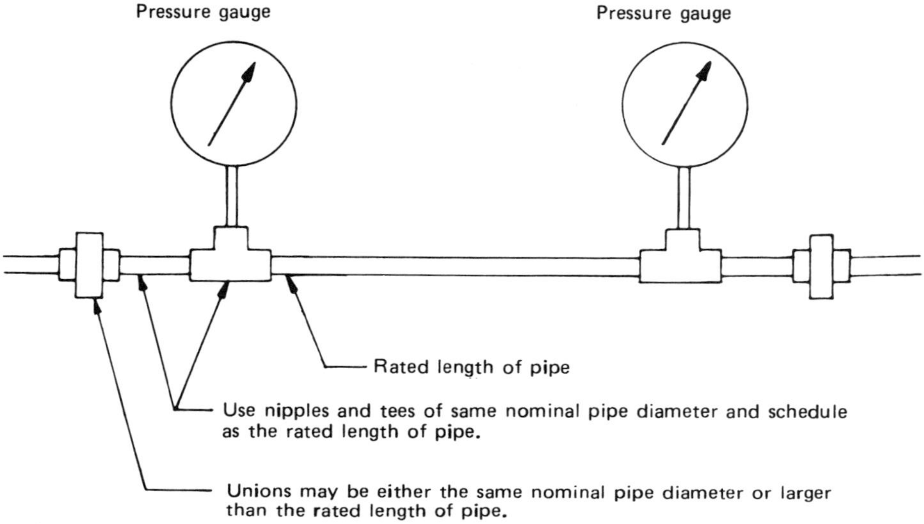

Figure B.3 Schematic of valveshift flow meter.

narrow range of air flow rates for moderate pressure differences such as 1 to 10 psi through the metering range. Because of normal variations in the dimensions of the pipe and fittings, as well as the presence of burrs in the cut pipe lengths, the error with this method can be expected to range between 10 and 20%. This concerns pipe sets made up and rated with estimated equivalent lengths of straight pipe in the plumbing supply house catalog. The technique of this estimating is indicated in Table B.1. Notice that only one tee is considered. Actually, it represents the two halves of the upstream and downstream pressure tap tees, which are part of the rated pipe section between the pressure taps. Figure B.1 can be used to convert the equivalent pipe lengths into flow factors.

Table B.1 shows an example of air flow rates at 1 and 10 psi pressure loss through the 1-ft.long pipe lengths between the tees with an upstream pressure of 80 psig. Notice that while these pipe sets encompass the range of 13 to 3040 cubic feet of air per minute, no pipe set can be eliminated without leaving a gap in the metering range. However, this is simply an example, and it

Table B.1 Dimensions and Flow Data of Pipe Sets

Nominal pipe diameter (in.)	Pipe length (ft)	Tee equivalent length (ft)	Total length (ft)	Flow factor	Air flow rate at 80 psig supply pressure (scfm)	
					1 psi loss	10 psi loss
1/8	1	0.48	1.48	0.84	13.1	41.1
1/4	1	0.65	1.65	1.75	27.3	85.8
3/8	1	0.88	1.88	3.95	61.7	193.5
1/2	1	1.11	2.11	6.75	105.1	331
3/4	1	1.4	2.4	13.5	211	662
1	1	1.8	2.8	23.8	372	1166
1 1/4	1	2.3	3.3	44	687	2155
1 1/2	1	2.7	3.7	62	967	3040

may be possible to alter the lengths of the pipes so that successive sets encompass the ranges more effectively. Actually, the basic thought behind the discussion of this method of flow metering is to explain how to work with whatever pipe lengths are available. It is suggested that the pipe lengths not be shortened to less than 1 ft, because this condition will then approach that for orifice flow, which is more aptly described with any of the standard orifice flow calculations.

If maximum accuracy is desired, calculate the flow factor of the pipe set from the actual internal dimensions of the pipe and tees [3], or determine the flow factor with a pressure loss and air flow rate test. Also, remove the cutting burrs at the entrance and exit of the nipples and the rated section of pipe. With reliable pressure gauges, this method should give results within 10% of the actual flow rates.

COMPONENT ORDER VERSUS SYSTEM ERROR

This effect is the result of the components' relative flow factors and the particular sequence of the components in the circuit. For example, we are aware that pressure loss calculations made for each component in succession will reflect the closest approximation to the real losses. However, this method is laborious, and it does not warn the designer of the presence of the critical pressure early in the calculations. On the other hand, the ability to combine components by merely adding together their equivalent straight pipe lengths, and then treating this total as a single restriction, greatly simplifies the pneumatic analysis. The relation of the air flow rate and the critical pressure can can then be conveniently evaluated, and if desired or necessary, the component capacities may be altered to provide for an adequate system.

Generally, the combining of the flow factors by the National Bureau of Stándards method is accomplished without consideration of the actual order of the components in the system. However this method gives results which are representative of the highest stagnation losses due to the least favorable sequence of the components in series. Therefore, any other arrangement of the same components can demonstrate actual flow factors up to 10% greater than calculated.

Table B.2 Data for Flow Factor Calculation Example

Component	P_1 (psia)	P_2 (psia)	Q (scfm)	Flow factor Q/P_1	Flow factor $\frac{Q}{1.153(P_1^2 - P_2^2)^{1/2}}$	Equivalent length of 1/2 in. NPD (ft)
1	95	47.5	100	1.053		90
2	47.5	23.75	100	2.105		23
3	23.75	14.7	100		4.65	4.7
					Total equivalent length:	117.7

An illustration of the above item might be as follows. Table B.2 shows the data.

As part of the example, suppose that three components in series are in a pneumatic circuit, as shown in Fig. B.4. Assume that 100 standard cubic feet of air per minute passes through these components. Now, if the upstream pressure to component 1 is 95 psia, and the downstream pressure is 47.5 psia, this component will be recognized as passing the maximum air flow to be expected, since the downstream absolute pressure is half of the upstream absolute pressure. Let us further assume that component 2 has a supply pressure of 47.5 psia, and the downstream pressure of 23.75 psia is half of the upstream pressure. Consequently, the

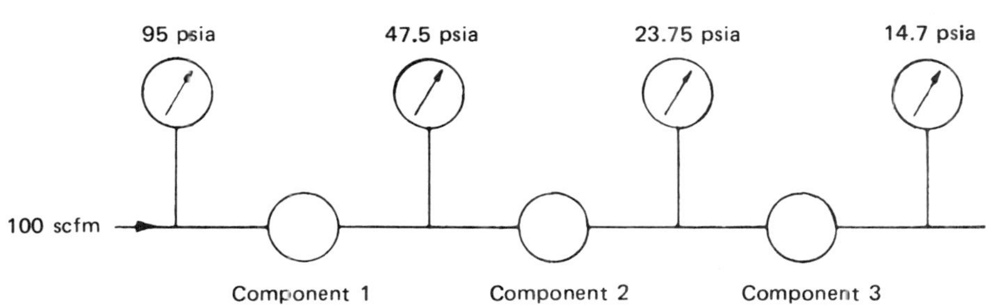

Figure B.4 Illustration for flow factor calculations.

maximum flow is 100 standard cubic feet of air per minute. Finally, component 3 with the upstream pressure of 23.75 psia has a downstream pressure of 14.7 psia, which is recognized as atmospheric pressure. Since the downstream pressure is not equal to the critical pressure of component 3, this unit is not passing its maximum air flow rate. However, the flow factor of this component was selected to pass the 100 standard cubic feet of air per minute at the remaining pressure loss.

By calculation with the National Bureau of Standards method, the three components in series in Figure B.4, with a total equivalent length of 117.7 ft of 1/2-in. nominal pipe diameter, will permit a maximum air flow rate of 86.6 standard cubic feet of air per minute. This can be compared with an actual test with the least favorable arrangement of the three components, which resulted in an air flow rate of 92 standard cubic feet of air per minute.

Theoretically, the optimum arrangement of components in series occurs when the flow factors of the successive components in series are related to the flow factor of the first component by the equation $2^{n-1}F_a$, where F_a is the flow factor of the upstream component and n is the component number counting from the upstream unit as number one. Therefore, the successive component flow factors would be F_a, $2F_a$, $4F_a$, $8F_a$, . . . , $2^{n-1}F_a$ as the optimum arrangement.

The above arrangement suggests that the otherwise optimum succession of components in series occurs when the components are situated with progressively increasing flow factors beginning with the upstream unit having the smallest flow factor and working in the downstream direction.

In situations similar to the above illustrations, the efforts to minimize the stagnation losses due to component sequence should not be expected to increase the actual rating of the system by more than 10%, and that this higher efficiency would be apparent mainly after a step-by-step circuit analysis. Still, an appreciation of this fact can aid in the decision to use increased capacities for the downstream units. Otherwise, the value of this review is mainly to illustrate the basic items that contribute to the maximum flow rate variations from the theoretical because of component arrangement.

REFERENCES

1. Slawsky, M. M., Lutzky, M., and Schmidlin, A. E., *Pressure Drop in Pneumatic Components and Systems.* Product Engineering Annual Handbook of Product Design (1955), J2–J5.
2. Riske, G., *Pressure Losses in Pneumatic Systems.* Product Engineering 30 (October 26, 1959), 50-54.
3. Riske, G., *Flow Factors.* Machine Design 32 (April 14, 1960), 154–159.

Index